Bioconversion Processes

Special Issue Editor
Christian Kennes

MDPI • Basel • Beijing • Wuhan • Barcelona • Belgrade

MDPI

Special Issue Editor
Christian Kennes
University of La Coruña
Spain

Editorial Office
MDPI
St. Alban-Anlage 66
Basel, Switzerland

This edition is a reprint of the Special Issue published online in the open access journal *Fermentation* (ISSN 2311-5637) from 2017–2018 (available at: http://www.mdpi.com/journal/fermentation/special_issues/bioconversion_processes).

For citation purposes, cite each article independently as indicated on the article page online and as indicated below:

Lastname, F.M.; Lastname, F.M. Article title. *Journal Name* **Year**, *Article number*, page range.

First Edition 2018

ISBN 978-3-03842-945-6 (Pbk)
ISBN 978-3-03842-946-3 (PDF)

Table of Contents

About the Special Issue Editor

Christian Kennes is full Professor of Chemical Engineering at the University of La Coruña (UDC) in Spain. He first undertook engineering studies and later obtained his Master and PhD degrees in Belgium. He belongs to the BIOENGIN ("Environmental Bioengineering and Quality Control") research group. He has been teaching subjects related to chemical and biochemical engineering, as well as environmental technology, wastewater treatment, waste gas treatment, and biorefinery processes. His main research topics presently focus on the removal of pollutants from water, air, and solid waste, mainly in bioreactors, as well as the fermentation and bioconversion of such pollutants and other renewable feedstocks into high value commercial products, including biofuels, biopolymers, and platform chemicals.

fermentation

MDPI

Editorial

Bioconversion Processes

Christian Kennes

Chemical Engineering Laboratory, Faculty of Sciences and Center for Advanced Scientific Research (CICA), University of La Coruña, Rúa da Fraga 10, E-15008 La Coruña, Spain; kennes@udc.es

Received: 10 March 2018; Accepted: 21 March 2018; Published: 23 March 2018

Keywords: anaerobic bacteria; biofuels; biomass; bioproducts; biorefinery; fungi; microalgae; solid waste; yeast; wastewater

---

Bioprocesses represent a promising and environmentally friendly option to replace the well-established chemical processes used nowadays for the production of platform chemicals, fuels, and other commercial products. Significant research is being performed to optimize bioconversion processes and biorefineries, which do already coexist, to some extent, with conventional refineries.

A range of different options and technologies are being studied and are presently available to obtain different useful end-products through bioprocesses. Many such processes focus on renewable resources, biomass, or pollutants as primary feedstocks. The latter avoid food–fuel competition, contrary to some other feedstocks considered in the past, and, sometimes, still today. This special issue offers some examples of interesting alternatives. Some suitable feedstocks include biomass [1,2], solid waste [3–5], sludge [6], wastewater [7,8], waste gases [9], or even byproducts, such as glycerol, from other biorefinery processes [10,11]. Several of those feedstocks and their corresponding bioconversion processes are addressed here. Some prime matters may need specific pre-treatments before undergoing microbial fermentation, such as those composed of complex polymeric materials, which first need to be converted to smaller or monomeric molecules in order to be accessible and metabolized by microorganisms [1,4,12]. Different types of microorganisms have been studied and can be used as biocatalysts, including pure or mixed cultures of aerobic and anaerobic bacteria [6,13], yeasts and fungi in general [1,3], as well as algae. The biocatalysts may be wild-type or engineered ones [10]. Direct application of enzymes can also be considered.

Bioconversion processes generally take place in bioreactors, which may be operated in batch, continuous, or semi-continuous mode, among others. Moreover, different bioreactor configurations may be suitable depending on the specific application. The technology may range from solid-phase bioconversion processes to gas-phase ones, besides aqueous phase bioprocesses. In any case, a given amount of moisture is generally needed, as this is required, in most cases, for optimal microbial activity. For any given feedstock, biocatalyst and bioreactor configuration and operating conditions will need to be optimized, in terms of aspects such as residence time in continuous processes, pH, or media composition (e.g., C/N ratio), as studied and reported in several manuscripts in this issue [3,10,14].

In conclusion, bioconversion processes and biorefineries are environmentally friendly alternatives to common chemical processes and conventional oil refineries. They allow the production of a wide range of products with cheap biocatalysts, usually under mild conditions. Additional intensive research is still needed in order to further optimize such processes.

Conflicts of Interest: The author declares no conflict of interest.

References

1. Xiu, S.; Bo Zhang, B.; Boakye-Boaten, N.A.; Shahbazi, A. Green Biorefinery of Giant Miscanthus for Growing Microalgae and Biofuel Production. *Fermentation* **2017**, *3*, 66. [CrossRef]

2. Nghiem, N.P.; O'Connor, J.P.; Hums, M.E. Integrated Process for Extraction of Wax as a Value-Added Co-Product and Improved Ethanol Production by Converting both Starch and Cellulosic Components in Sorghum Grains. *Fermentation* **2018**, *4*, 12. [CrossRef]

3. Mahboubi, A.; Ferreira, J.A.; Taherzadeh, M.J.; Lennartsson, P.R. Production of Fungal Biomass for Feed, Fatty Acids, and Glycerol by *Aspergillus oryzae* from Fat-Rich Dairy Substrates. *Fermentation* **2017**, *3*, 48. [CrossRef]

4. Velasco, D.; Senit, J.J.; de la Torre, I.; Santos, T.M.; Yustos, P.; Santos, V.E.; Ladero, M. Optimization of the Enzymatic Saccharification Process of Milled Orange Wastes. *Fermentation* **2017**, *3*, 37. [CrossRef]

5. Chalima, A.; Oliver, L.; de Castro, L.F.; Karnaouri, A.; Dietrich, T.; Topakas, E. Utilization of Volatile Fatty Acids from Microalgae for the Production of High Added Value Compounds. *Fermentation* **2017**, *3*, 54. [CrossRef]

6. Alrawashdeh, K.A.B.; Pugliese, A.; Slopiecka, K.; Pistolesi, V.; Massoli, S.; Bartocci, P.; Bidini, G.; Fantozzi, F. Codigestion of Untreated and Treated Sewage Sludge with the Organic Fraction of Municipal Solid Wastes. *Fermentation* **2017**, *3*, 35. [CrossRef]

7. Souza Filho, P.F.; Brancoli, P.; Bolton, K.; Zamani, A.; Taherzadeh, M.J. Techno-Economic and Life Cycle Assessment of Wastewater Management from Potato Starch Production: Present Status and Alternative Biotreatments. *Fermentation* **2017**, *3*, 56. [CrossRef]

8. Ben, M.; Kennes, C.; Veiga, M.C. Optimization of polyhydroxyalkanoate storage using mixed cultures and brewery wastewater. *J. Chem. Technol. Biotechnol.* **2016**, *91*, 2817–2826. [CrossRef]

9. Fernández-Naveira, Á.; Veiga, M.C.; Kennes, C. H-B-E (hexanol-butanol-ethanol) fermentation for the production of higher alcohols from syngas/waste gas. *J. Chem. Technol. Biotechnol.* **2017**, *92*, 712–731. [CrossRef]

10. Abghari, A.; Chen, S. Engineering *Yarrowia lipolytica* for Enhanced Production of Lipid and Citric Acid. *Fermentation* **2017**, *3*, 34. [CrossRef]

11. Matsakas, L.; Hrůzová, K.; Ulrika Rova, U.; Christakopoulos, P. Biological Production of 3-Hydroxypropionic Acid: An Update on the Current Status. *Fermentation* **2018**, *4*, 13. [CrossRef]

12. Kennes, D.; Abubackar, H.N.; Diaz, M.; Veiga, M.C.; Kennes, C. Bioethanol production from biomass: Carbohydrate vs. syngas fermentation. *J. Chem. Technol. Biotechnol.* **2016**, *91*, 304–317. [CrossRef]

13. Fernández-Naveira, Á.; Abubackar, H.N.; Veiga, M.C.; Kennes, C. Production of chemicals from C1 gases (CO, CO_2) by *Clostridium carboxidivorans*. *World J. Microbiol. Biotechnol.* **2016**, *33*, 43. [CrossRef] [PubMed]

14. Ginésy, M.; Rusanova-Naydenova, D.; Rova, U. Tuning of the Carbon-to-Nitrogen Ratio for the Production of L-Arginine by *Escherichia coli*. *Fermentation* **2017**, *3*, 60. [CrossRef]

fermentation

MDPI

Article

Engineering *Yarrowia lipolytica* for Enhanced Production of Lipid and Citric Acid

Ali Abghari * and Shulin Chen *

Department of Biological Systems Engineering, Bioprocessing and Bioproducts Engineering Laboratory, Washington State University, Pullman, WA 99163, USA
* Correspondence: ali.abghari@wsu.edu (A.A.); chens@wsu.edu (S.C.);
 Tel.: +1-509-336-9227 (A.A.); +1-509-335-3743 (S.C.)

Received: 2 May 2017; Accepted: 12 July 2017; Published: 17 July 2017

Abstract: Increasing demand for plant oil for food, feed, and fuel production has led to food-fuel competition, higher plant lipid cost, and more need for agricultural land. On the other hand, the growing global production of biodiesel has increased the production of glycerol as a by-product. Efficient utilization of this by-product can reduce biodiesel production costs. We engineered *Yarrowia lipolytica* (*Y. lipolytica*) at various metabolic levels of lipid biosynthesis, degradation, and regulation for enhanced lipid and citric acid production. We used a one-step double gene knock-in and site-specific gene knock-out strategy. The resulting final strain combines the overexpression of homologous *DGA1* and *DGA2* in a *POX*-deleted background, and deletion of the *SNF1* lipid regulator. This increased lipid and citric acid production in the strain under nitrogen-limiting conditions (C/N molar ratio of 60). The engineered strain constitutively accumulated lipid at a titer of more than 4.8 g/L with a lipid content of 53% of dry cell weight (DCW). The secreted citric acid reached a yield of 0.75 g/g (up to ~45 g/L) from pure glycerol in 3 days of batch fermentation using a 1-L bioreactor. This yeast cell factory was capable of simultaneous lipid accumulation and citric acid secretion. It can be used in fed-batch or continuous bioprocessing for citric acid recovery from the supernatant, along with lipid extraction from the harvested biomass.

Keywords: *Yarrowia lipolytica*; microbial lipid; citric acid; glycerol; genetic and metabolic engineering; fermentation; leucine metabolism and biosynthesis; bioconversion

1. Introduction

Volatility of energy price and concerns over climate change have motivated efforts to explore alternative approaches for production of fuels and chemicals. Microbial fermentation of low-value biomass is a promising strategy for sustainable production of these compounds. Single-cell-oil (SCO), for example, is of great interest to the food, nutraceuticals, and biodiesel industries. Oleaginous organisms such as fungi, yeasts, and algae can accumulate oil beyond 20% of their biomass under appropriate cultivation conditions [1]. The application of oleaginous yeasts as a lipid-producing platform offers many advantages. These include feedstock flexibility, higher sustainability, shorter life cycles, easy cultivation and handling, robustness against contamination, seasonal independence, and lower net greenhouse gas emissions [1,2].

Industrial-scale production of SCO is challenging due to large volumes and low profit margins [3]. Technological and cellular-level improvements are required to reduce processing costs and achieve higher productivity with wider range of low-value substrates [4]. Prior to genetic modification, the lipid content of a wild-type *Y. lipolytica* strain rarely reaches 20% DCW [5]. Therefore, metabolic engineering is necessary to improve lipid productivity. Additionally, the production of other value-added co-products and exploration of zero-cost waste or by-product streams such as glycerol, as feedstock, for yeast SCO production is recommended [6].

Plant-based production of biodiesel is anticipated to reach 30×10^6 t. in 2021. Since 1 kg glycerol is produced per 10 kg of biodiesel, this would generate 3×10^6 t. glycerol as by-product [7]. Valorization of glycerol for producing SCO or other higher added-value compounds offsets the costs of biodiesel, reduces glycerol surplus, and favors the viability of SCO bioprocess.

Much research has focused on the oleaginous yeast *Y. lipolytica*, a known model non-conventional yeast, to produce and/or secrete various oleochemicals and recombinant proteins [8–10]. This platform is commonly considered for production of lipid, citric acid, as well as oleochemicals derived from acetyl-CoA and fatty acid [11,12]. Although *Y. lipolytica* and *Aspergillus niger* are major producers of citric acid [13], the former is more resistant to metals and offers more environmentally friendly process [14]. *Y. lipolytica* can release both citric acid, at higher concentration, and its isomer isocitric acid at lower concentration. This ratio depends on the feedstock [15]. For example, Morgunov et al., fed this yeast with pure and raw glycerol in a fed-batch cultivation for citric acid production. They reported a citric acid/isocitric acid ratio of 21 to 25, with isocitric acid represented up to 5% [16]. While citric acid is an extracellular metabolite and is secreted into the culture medium, lipid is intracellularly stored in the form of triglycerides (TAG) in this oleaginous yeast. TAG does not have lipotoxicity on the cells as free fatty acids do [17], and can accommodate essential and non-essential fatty acids and precursors for dynamic cell maintenance. *Y. lipolytica* has also shown promise in the bioconversion of glycerol as renewable feedstock to various compounds [18], including lipid [19–24] and citric acid [13,16]. This yeast can efficiently utilize glycerol and prefers it over many other carbon sources [25]. It also has a similar rate of lipid production when fed with pure or crude forms of glycerol [26]. Therefore, this yeast can play a dual role in upstream and downstream processes of biodiesel industries by producing microbial lipid and other valuable pharmaceuticals from glycerol [19].

In this study, we aimed to engineer *Y. lipolytica* to enhance lipid and citric acid production from pure glycerol. We took advantage of the one-step gene knock in/out for targeted integration and overexpression of key TAG synthesizing genes, followed by deletion of *SNF1* gene in the *POX* deleted strain. This strategy served constitutive diversion of carbon flux into the neutral lipid and citric acid in nitrogen-limited glycerol-based media supplemented with leucine. We also examined the effect of leucine supplementation or *LEU2* expression on metabolite production and biomass generation. We cultivated engineered *Y. lipolytica* strains in a shake flask and then performed batch cultivation in a 1-L bioreactor under well-controlled conditions to enhance lipid and citric acid productivity.

2. Materials and Methods

2.1. Strains and Culture Condition

Table 1 describes the recombinant *Y. lipolytica* strains that were derived from the citric acid producer strain H222 (wild-type German strain) [27]. *Escherichia coli* top 10 was used to develop vectors. Ampicillin was added to the Luria-Bertani (LB) broth medium at concentration of 100 µg/mL according to standard protocols [28].

Table 1. *Yarrowia* strains used in this study.

Y. Lipolytica Strain Names	Strain Genotypes	Gene Configurations	Reference
H222 (H)	MatA mating type		[27]
H222ΔP *leu*⁺ *ura*[−] (HP-U)	MATA *ura3-302::SUC2* ΔPOX1-6		[27]
H222ΔP *leu*⁺ *ura*⁺ (HP)	HP-U, ΔPOX3::URA3	*loxR-URA3-loxP* flanked by *POX3* homologous up/down stream sequences	This study
H222ΔP ΔL + DGA1 DGA2 *leu*[−] *ura*⁺ (HPDD)	HP, ΔLEU2 + DGA1 + DGA2::URA3	*loxR-URA3-loxP* flanked by *LEU2* homologous upstream and pFBA-DGA1-tLip1 pTEF-DGA2-tXPR2 *LEU2* homologous downstream sequences	This study
H222ΔP ΔL + DGA1 DGA2 ΔSNF1 *leu*[−] *ura*⁺ (HPDDS)	HPDD, ΔSNF1::URA3	*loxR-URA3-loxP* flanked by *SNF1* up/down homologous stream sequences	This study

5

Synthetic defined media containing 6.7 g/L Yeast Nitrogen Base (YNB) w/ammonium sulfate w/o amino acids (Becton, Dickson, and company), 20 g/L glucose, and a drop-out synthetic mix minus uracil (-Ura) or minus leucine (-Leu) (US Biological) were used for the selection of knock out/in strains. The uracil auxotrophic strains were obtained by growing in YNB-Leu liquid medium with the expression of Cre recombinase. Seed culture preparation was carried out using the synthetic defined medium devoid of uracil (YNB-Ura). A rich medium (YPD) was prepared with 20 g/L glucose, 20 g/L bacto peptone (BD), and 10 g/L bacto yeast extract (BD), and was used for non-selective propagation of strains. The YNB-Ura and YNB-Leu media were buffered with a 50 mM sodium phosphate buffer, pH 6.8, to determine the effects of leucine supplementation and *LEU2* expression on biomass and metabolite production. For solid media, 20 g/L agar (US Biological, Swampscott, MA, USA) was added.

For lipid production in the shake flask and bioreactor, previous data on glycerol based fermentation media was taken into account, followed by some modifications [19,29]. The medium was formulated as follows: 1.5 g/L yeast extract (BD), 1.5 g/L $MgSO_4 \cdot 7H_2O$, 7 g/L KH_2PO_4, 2.5 g/L Na_2HPO_4, 0.15 g/L $CaCl_2 \cdot 2H_2O$, 0.15 g/L $FeCl_3 \cdot 6H_2O$, 0.02 g/L $ZnSO_4 \cdot 7H_2O$, 0.06 g/L $MnSO_4 \cdot H_2O$, 0.1 mg/L $CoCl_2 \cdot 6H_2O$, and 0.04 mg/L $CuSO_4 \cdot 5H_2O$. Prior research suggested for this yeast, glycerol concentration should range from 52 to 112 g/L for bioconversion of glycerol to biomass and lipid [21]. In the batch cultivations of this study, glycerol solution was separately sterilized and added to the flasks to reach an initial concentration of 60 ± 2 g/L. The carbon to nitrogen ratio (C/N) was adjusted to 60 for all production media using pure glycerol (J.T. baker) and 1.1 g/L $(NH4)_2SO_4$ as major carbon and nitrogen sources, respectively. Leucine was added to production media in shake flask and bioreacotr at a concentration of 100 mg/L (Teknova) to compensate for *LEU2* deletion in the HPDD and HPDDS strains. Shake flask cultivations were performed in 250 mL Erlenmeyer flasks containing 50 mL of the medium at an agitation rate of 180 ± 5 rpm and temperature of 28 ± 1 °C. Colonies from solid YNB-Ura plates were precultured in the selective defined media. Exponentially growing cells were harvested by centrifuge, washed and then resuspended in water. They were subsequently inoculated into the production medium to reach an initial optical density (OD_{600}) of 0.1.

2.2. Batch Fermentation

Batch cultivation was carried out in a 1-L benchtop fermenter, BioFlo 110 (New Brunswick Scientific, Enfield, CT, USA). A single colony of *Y. lipolytica* grown on DOB-Ura was transferred into the YNB-Ura broth. Cells from 100 mL 24 h shake flask pre-culture were harvested by centrifugation at 12,000 rpm, washed twice with water and inoculated into 700 mL of the fermentation medium (with C/N 60) to reach an initial OD_{600} of ~0.3. The temperature was kept at 28 °C, and the pH was controlled not to drop below 2.5, using 1 M NaOH. Dissolved oxygen was maintained at 25% until peak biomass was attained (from 48 h to ~72 h). This was achieved by cascading with agitation ranging from 250 to 800 rpm, and by supplying sterile, filtered air at flow rate of 2 vvm. The dissolved oxygen and airflow rate were later decreased to ~5% and 0.5 vvm, respectively, near the end of the 5-day fermentation. The fermenter experiments were performed in duplicate. Samples with the volume of 25 mL were taken daily. An antifoam Y-30 emulsion (Sigma-Aldrich, St. Louis, MO, USA) solution was prepared at a concentration of 5%, and was periodically added to control the foam level.

2.3. Genetic Techniques

Standard molecular biology techniques were used to construct the vectors [28]. Table 2 presents all plasmids and their functions (See Supplementary Materials for plasmid maps).

Table 2. Vectors used in this study.

Vector Names	Features
Cre-recombinase (CR)	Shuttle vector carrying leucine marker, Cre recombinase flanked by TEFin promoter and Xpr2 terminator
pGR12 (L)	Shuttle empty vector carrying leucine marker, FBA promoter and lip1 terminator, used for study of leucine biosynthesis
POX3 Ura (PU)	Uracil selection marker flanked by *POX3* upstream and downstream homologous sequences, used for construction of HP strain
LEU2 Ura (LU)	Uracil selection marker flanked by *LEU2* upstream and downstream homologous sequences, used for construction of LDD vector
SNF1 Ura (SU)	Uracil selection marker flanked by *SNF1* upstream and downstream homologous sequences, used for construction of HPDDS strain
pGR12 DGA1 (D1)	Single gene centromeric shuttle replicative vector with leucine selection marker, *DGA1* gene cloned between FBA promoter and lip1 terminator, used for double gene expression cassette construction
pJN44 DGA2 (D2)	Single gene centromeric shuttle replicative vector with leucine selection marker, *DGA2* gene cloned between TEFin promoter and xpr2 terminator, used for double gene expression cassette construction
DGA1 DGA2 (DD)	Double gene centromeric shuttle replicative vector with leucine selection marker, used for construction of LDD vector
LEU2 DGA1 DGA2 (LDD)	Uracil selection marker flanked by *LEU2* homologous upstream sequence and combination of double gene expression cassettes and *LEU2* homologous downstream sequence, used for construction of HPDD strain

Construction of the double gene expression cassette was carried out by amplification of diacylglycerol acyltransferases *DGA1* (YALI0E32769g) and *DGA2* (YALI0D07986g) gene segments using the Q5 high fidelity DNA polymerase (New England Biolabs, Ipswich, MA, USA) and gDNA from Po1f (ATCC MYA-2613) as the template with the primers listed in Table 3. The *DGA1* and *DGA2* amplicons were individually digested and inserted into *Y. lipolytica* plasmid pGR12 (PFBA-Tlip1) and pJN44 (PTEFin-Txpr2), respectively. The segment of PTEFin-*DGA2*-Txpr2 was obtained by digestion with XbaI and SpeI and then, recovered from the gel. Then it was inserted into SpeI and Fast Alkaline Phosphatase digested *DGA1*–pGR12 plasmid.

Plasmids for gene knock-out contained the uracil selection marker surrounded by LoxP sites. For knock-out plasmid constructions, the 0.6–1.1 kb 5'- and 3'-flanking regions of the *Y. lipolytica LEU2*, *SNF1*, and *POX3* genes were amplified with the primers listed in Table 3. The amplicons were digested, purified, and inserted into the upstream and downstream of the uracil marker. The double gene expression cassette segment underwent double digestion with XbaI and SpeI followed by gel recovery for subsequent insertion into SpeI and Fast Alkaline Phosphatase digested *LEU2* knock-out plasmid (LU), which was used in one step knock in/out.

Targeted gene knock in/out was achieved by transformation of the linearized vectors containing homologous upstream and downstream sequences. The linearized vectors consisted of NdeI-digested PU, ApaI-digested SU, and NdeI-digested LDD plasmids. Transformation was performed using the Zymogen Frozen EZ yeast transformation kit II (Zymo Research, Irvine, CA, USA), in compliance with the manufacturer's protocol. The *loxR–URA3–loxP* modules were rescued for subsequent genetic modification by the LoxP-Cre system as previously reported [30]. Gene deletions and expression cassette insertion were confirmed by Polymerase chain reaction (PCR) using the primers listed in Table 3.

Table 3. PCR primers used in this study.

No.	Name	Sequence (5′—›3′, Underlined Restriction Site)
1	*POX3* up F ApaI	CTATAGGGCCCCTGGGCTGTTCGGTCGA
2	*POX3* up R XbaI	GATCCTCTAGAAGGACGCACAACGCC
3	*POX3* down F SpeI	CTGGACTAGTCGCTCCCATTGGAAACTACGA
4	*POX3* down R NdeI	CCTCACATATGTCTCTTCGCTGTGGTCTAGG
5	*POX3* F Ura	GTCTCTACTTGTAGTTCTGTAGACAGACT
6	*POX3* Ura R	GAAGAATGTATCGTCAAAGTGATCCAAG
7	*POX3* Ura F	TGACTTGTGTATGACTTATTCTCAACTACA
8	*POX3* R Ura	AGATGCGTGATAGATTACTTGGATTTAGT
9	*DGA1* F HindIII	GAGCGAAAGCTTATGACTATCGACTCACAATACTACAAGT
10	*DGA1* R SalI	GTTCAAGTCGACTTACTCAATCATTCGGAACTCTGGG
11	*DGA2* F HindIII	GCAAGGAAGCTTATGGAAGTCCGACGACGA
12	*DGA2* R PstI	ATGCTACTGCAGCTACTGGTTCTGCTTGTAGTTGT
13	*LEU2* up F ApaI	CTATAGGGCCC ACCGGCAAGATCTCGTTAAGACAC
14	*LEU2* up R XbaI	GATCCTCTAGATGTGTGTGGTTGTATGTGTGATGTGG
15	*LEU2* down F SpeI	CTGGACTAGTCTCTATAAAAAGGGCCCAGCCCTG
16	*LEU2* down R NdeI	CCTCACATATG GACAGCCTTGACAACTTGGTTGTTG
17	*LEU2* F Ura	TACAGTTGTAACTATGGTGCTTATCTGGG
18	*LEU2* Ura R	CCTTGGGAACCACCACCGT
19	*LEU2* Ura F	ACTTCCTGGAGGCAGAAGAACTT
20	*LEU2* R Ura	ATAGCAAATTTAGTCGTCGAGAAAGGGTC
21	*SNF1* up F ApaI	CAATTGGGCCCGTGATCAAAGCATGAGATACTGTCAAGG
22	*SNF1* up R XbaI	GATCCTCTAGAGAGGTGGTGGAAGGAGTGGTATGTAGTC
23	*SNF1* down F SpeI	CTGGACTAGT TCATTAATACGTTTCCCTGGTG
24	*SNF1* down R NdeI	CCTCACATATGGGGAATTCGTGCAGAAGAACA
25	*SNF1* F Ura	GCGGGAAATCAAGATTGAGA
26	*SNF1* Ura R	CGGTCCATTTCTCACCAACT
27	*SNF1* Ura F	CCTGGAGGCAGAAGAACTTG
28	*SNF1* R Ura	ACTACTGGCGGACTTTGTGG

The plasmids were constructed using standard restriction digestion cloning with FastDigest restriction enzymes (Thermo Fisher Scientific, Waltham, MA, USA). Yeast genomic DNA was prepared for PCR amplification and verification as described previously [31]. The DNA products of PCR and digestion were purified with the clean and concentrator-5 Kit (Zymo Research). DNA fragments were recovered from agarose gels with a GeneJET Gel Extraction Kit (Thermo Fisher Scientific).

2.4. Analytical Methods

2.4.1. Dry Biomass

Seven-milliliter samples were collected daily. Five-milliliter samples were centrifuged for 5 min at 13,300 rpm. The cell pellet was washed first with saline (0.9% NaCl solution) and then with distilled water. The biomass yield was determined gravimetrically after the samples were dried at 105 °C until a consistent weight was reached. This was expressed in grams of dry cell weight per liter (g DCW/L).

2.4.2. Glycerol and Citric Acid Concentrations

Concentrations of glycerol and citric acid in fermentation broth were analyzed by varian Pro Star 230 high-performance liquid chromatography (HPLC) using an Aminex HPX- 87H column. Samples were centrifuged and supernatants were filtered using 0.22 μm pore-size membranes (Simsii, Inc., Irvine, CA, USA). Subsequently, they were eluted with 5 mM H_2SO_4 at a flow rate of 0.6 mL/min and 65 °C. Signals were detected by refractive index (RI) and UV (210 nm) detectors. Standards were used for identification and quantification of the glycerol and citric acid. This method was not able to distinguish between citric acid and its isomer isocitric acid. Thus, the sum of their concentrations was determined.

2.4.3. Qualitative and Quantitative Analysis of Lipids

Total lipid extraction and transesterification were carried out according to the procedure described previously by O'Fallon et al., 2007 [32]. Fatty acid methyl esters (FAME) were prepared in hexane and analyzed by gas chromatography (GC). This analysis was performed using an Agilent 7890A gas chromatography instrument coupled with a flame-ionization detector (FID) and a FAMEWAX column (30 m × 320 μm × 0.25 μm) (Restek Corporation, Bellefonte, PA, USA). The injection temperature and volume was set at 250 °C and 1 μL, respectively. The injection was performed with a split mode (ratio 20:1). The oven was initially 190 °C, and was increased to 240 °C at a rate of 5 °C min^{-1}. This was maintained at the final temperature for 20 min. The FID temperature was 250 °C. FAME standards were used to identify the fatty acid peaks in the chromatograms. The (0.5 mg/mL) tridecanoic acid (C13:0) (Sigma-Aldrich, St. Louis, MO, USA) solution in methanol was used as the internal standard to quantify the fatty acids. The total lipid titer and content was reported as g/L and percentage of the DCW, respectively. The supernatant was analyzed for possible extracellular lipid extraction.

3. Results

3.1. Comparative Time-Course Study

In this research, we constructed several strains through overexpressing key TAG-synthesizing genes and deleting the key negative regulator of the *de novo* fatty acid biosynthesis pathway. Specifically, the double gene expression cassette of *DGA1* and *DGA2* was integrated into *LEU2* locus of the ΔPOX1-6 HP strain to improve lipid synthesis and generate the HPDD strain. The deletion of *SNF1* was combined into ΔPOX1-6, ΔLEU2 DGA1p and DGA1p overexpression background to construct the HPDDS strain for creating potential synergy in carbon dedication to lipid and citric acid production. All of our strains were phototrophic for uracil. Pure glycerol was used as a carbon source at an initial concentration of 60 g/L under nitrogen-limiting conditions (C/N = 60). The following section presents data from the comparative time course study of feedstock consumption and the production of biomass, citric acid, and lipid by four strains. For this purpose, we collected samples at one-day intervals for six days. Although we present related data and previous findings from the literature, accurate comparison between the results and those of previous research is only possible when all variables are taken into account, including strain types, cultivation conditions, and genetic engineering strategies.

3.1.1. Glycerol Consumption

The comparative study of glycerol consumption by four strains during the 6-day fermentation was conducted. It can be seen in Figure 1 that the H and HP strains utilized almost all of the glycerol during the 6-day fermentation. However, 1–3 g/L glycerol remained from both the HPDD and HPDDS strains during the same period. The lower glycerol consumption rate may be due to a lower cell biomass level consuming the feedstock, lack of *LEU2* expression, and some metabolic perturbations caused by our genetic modification. It is noticeable that the genetic background of the strains affects the diversion of glycerol metabolism into specific pathways and outcomes. For instance, in the wild-type strain, the feedstock was used for more biomass production and corresponding cell maintenance, while in the HPDD and HPDDS strains, a higher portion of the feedstock was spent on lipid and citric acid production.

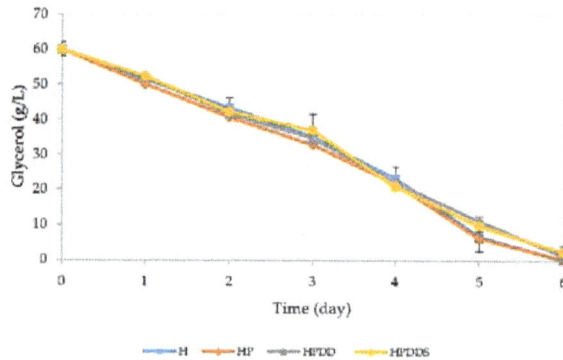

Figure 1. Comparative glycerol consumption by four strains during 6-day shake flask cultivation at $28 \pm 1\ °C$ under nitrogen limiting conditions (C/N = 60). Error bars represent standard deviation of n = 3. H: H222 wild-type strain, HP: H222 ΔPOX1-6, HPDD: H222 ΔPOX1-6 ΔLEU2 +DGA1 DGA2, HPDDS: H222 ΔPOX1-6 ΔLEU2 +DGA1 DGA2 ΔSNF1.

3.1.2. Biomass Production

The results of biomass production from 6-day shake flask cultivations for all four strains are summarized in Figure 2.

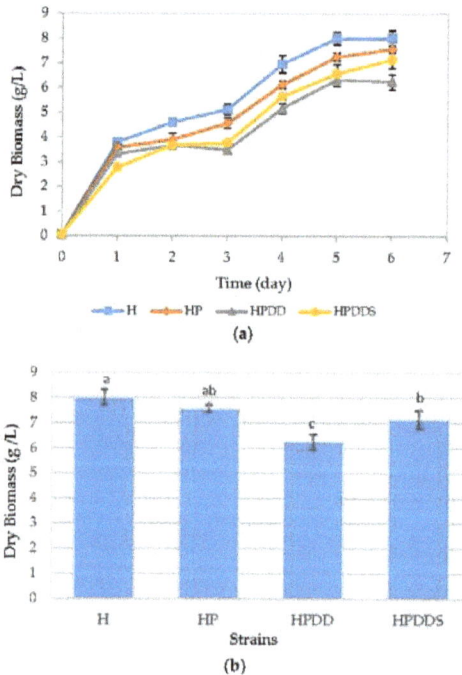

(a)

(b)

Figure 2. Comparative biomass production by four strains during 6-day shake flask cultivation (**a**) and on the last day (**b**) at $28 \pm 1\ °C$ under nitrogen limiting conditions (C/N = 60). Error bars represent standard deviation of n = 3. H: H222 wild-type strain, HP: H222 ΔPOX1-6, HPDD: H222 ΔPOX1-6 ΔLEU2 + DGA1 DGA2, HPDDS: H222 ΔPOX1-6 ΔLEU2 + DGA1 DGA2 ΔSNF1. abc columns with dissimilar letters at the top are significantly different (p < 0.05).

The yeast biomass was produced by four strains during six days of shake flask cultivations. As shown in Figure 2, the wild-type strain produced the highest level of the dried yeast biomass (about 8 g/L) under the study conditions. The deletion of *POX* genes slightly affected biomass formation, while simultaneous *LEU2* deletion and *DGA1, 2* overexpression led to a significant ($p < 0.05$) reduction of biomass to 6.3 g/L. This loss was recovered in part by a higher lipid accumulation caused by *SNF1* gene deletion. The engineered strain HPDDS formed 7.15 g/L of biomass during the six days of incubation.

3.1.3. Citric Acid Production

The results of the comparative time-course study performed on citric acid production from shake flask cultivations for all four strains are illustrated in Figure 3. Considering that citric acid was unstable in both the H and HP cultures, their maximum peaks were taken into account for statistical analysis.

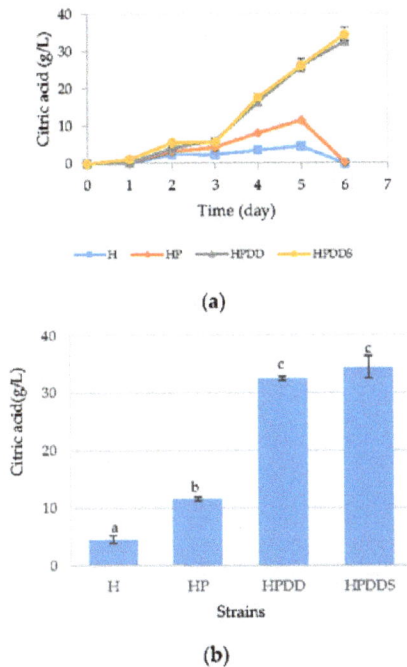

(a)

(b)

Figure 3. Comparative citric acid production by four strains during 6-day shake flask cultivation (a) and on the last day (b) at 28 ± 1 °C under nitrogen limiting conditions (C/N = 60). Error bars represent standard deviation of $n = 3$. H: H222 wild-type strain, HP: H222 $\Delta POX1$-6, HPDD: H222 $\Delta POX1$-6 $\Delta LEU2+DGA1$ $DGA2$, HPDDS: H222 $\Delta POX1$-6 $\Delta LEU2$ + $DGA1$ $DGA2$ $\Delta SNF1$. abc Columns with dissimilar letters at the top are significantly different ($p < 0.05$).

Our method could not distinguish between two isomers, citric acid and isocitric acid. Therefore, our reported concentration corresponds to the sum of these two acids. The results shown in Figure 3 indicate that all strains produced citric acid (a by-product of lipid biosynthesis) at different levels. It is interesting to note that the citric acid production was followed by citric acid degradation by the H and HP strains due to the exhaustion of glycerol, an extracellular carbon supply. In fact, *Y. lipolytica* is not only capable of citric acid production, but also use of it as a carbon and energy source [33]. Both the HPDD and HPDDS strains produced significantly ($p < 0.05$) more citric acid, ranging from 32 to 35 g/L, as the by-product of lipid biosynthesis. Consumption of citric acid was not observed for these two strains. This can be due to the availability of glycerol as a substrate during the 6-day shake flask fermentation. The HPDDS strain devoid of *SNF1* produced the highest level of citric acid during this period under nitrogen-limiting conditions. The maximum peak of citric acid was obtained at the end of incubation for the resting cells when the final pH was in the range of 2.3–2.5. One study suggested that the citric acid production occurs mainly during the stationary phase and is minimal at pH 3.0 [34]. Subsequently, we selected the best citric acid-producing strain, HPDDS, for further studies in the bioreactor.

3.1.4. Lipid Production

The results of lipid production by all four strains in the 6-day shake flask cultures are presented in Figure 4.

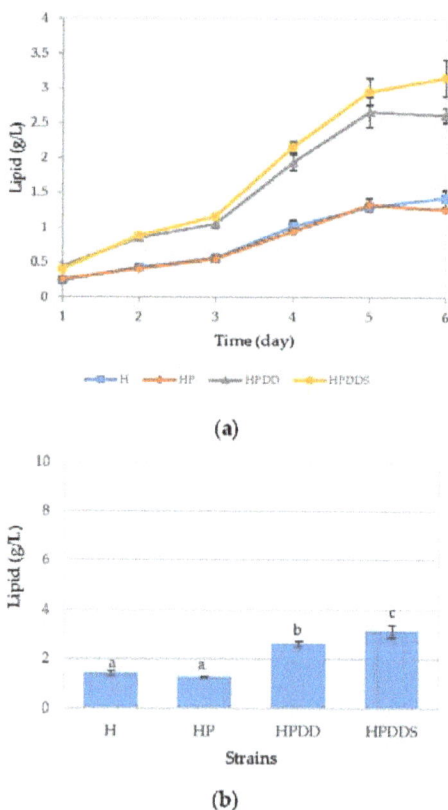

(a)

(b)

Figure 4. Comparative lipid production by four strains during 6-day shake flask cultivation (**a**) and on the last day (**b**) at 28 ± 1 °C under nitrogen limiting conditions (C/N = 60). Error bars represent standard deviation of *n* = 3. H: H222 wild-type strain, HP: H222 Δ*POX1-6*, HPDD: H222 Δ*POX1-6* Δ*LEU2* + *DGA1 DGA2*, HPDDS: H222 Δ*POX1-6* Δ*LEU2* + *DGA1 DGA2* Δ*SNF1*. abc columns with dissimilar letters at the top are significantly different (*p* < 0.05).

Lipid accumulation in the wild-type strain and the strain with the inactive β-oxidation degradation pathway was limited to 1.3–1.4 g/L, representing 17 to 18% of DCW under the nitrogen-limiting conditions. This observation accords with the low level of lipid accumulation (less than 1 g/L accounting for 3 to 20% of DCW) obtained by growing *Y. lipolytica* on biodiesel-derived glycerol under nitrogen-limiting conditions [35]. Lipid content can be enhanced through optimization of culture conditions [20] or through genetic manipulation. Our genetic engineering significantly ($p < 0.05$) increased the total fatty acid content to 2.6 g/L (42% of DCW) in the HPDD strain and to 3.15 g/L (44% of DCW) in the HPDDS strain in the 6-day shake flask cultivations. We observed an improvement of 2.47-fold in lipid content over the wild-type strain. The variation of lipid content percentages among all four strains is also notable (see Figure 5).

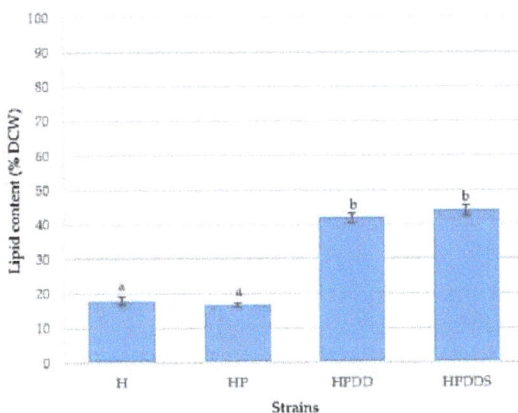

Figure 5. Comparative lipid content of four strains at the end of 6-day shake flask cultivation at $28 \pm 1\,^{\circ}$C under nitrogen limiting conditions (C/N = 60). Error bars represent standard deviation of $n = 3$. H: H222 wild-type strain, HP: H222 $\Delta POX1$-6, HPDD: H222 $\Delta POX1$-6 $\Delta Leu2$+$DGA1$ $DGA2$, HPDDS: H222 $\Delta POX1$-6 $\Delta Leu2$ + $DGA1$ $DGA2$ $\Delta SNF1$. ab columns with dissimilar letters at the top are significantly different ($p < 0.05$).

In comparison, another study coupled *DGA2* overexpression with *SNF1* deletion, resulting in a lipogenic phenotype for an engineered *Y. lipolytica* with a lipid content of over 76% using acetate as a carbon source [36]. Our titers are higher than those reported by Poli et al., who achieved 4.9 g/L biomass and 1.48 g/L lipid (30% of DCW) by growing *Y. lipolytica* QU21 on 100 g/L glycerol and NH4SO4 as a nitrogen source in a shake flask study [26]. Figure 5 indicates that HPDDS is the best lipid-producing strain compared to other strains under our study conditions (C/N of 60 and leucine supplementation of 100 mg/L). However, the lipid content of this strain is not significantly different ($p > 0.05$) from the HPDD strain. The HPDDS strain showed an increase in the lipid production of over 120% compared to the wild-type strain. Figure 6 presents the major fatty acid content of the strains. We optimized our GC-FID method for analysis of the fatty acid content. However, our final method and the corresponding GC-FID column could not efficiently separate C18 from C18:1 fatty acid. Therefore, we reported the sum of these two fatty acids together, with an approximate ratio of C18:1/C18 based on their corresponding peak heights.

The analysis of the fatty acid profile among different strains is presented in Figure 6. Oleic acid (C18:1) and palmitic acid (C16) were the predominant fatty acids, with a concentration that varied by strains. Oleic acid had the highest concentration, ranging from 44% to 51% in the strains. The predominance of oleic acid is in accordance with the 47% and 59% oleic acid reported by Papanikolaou et al. (2013) for the wild and engineered strains grown on 90 g/L of glycerol [37]. The other minor fatty acids, stearic acid (C18), linoleic acid (C18:2), and palmitoleic acid (C16:1), underwent smaller changes as a result of our genetic modification.

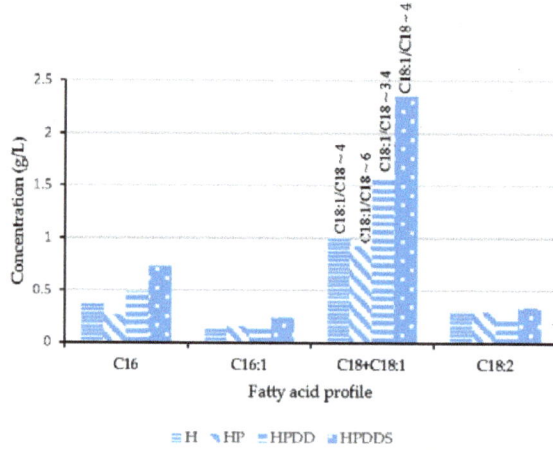

Figure 6. Comparative fatty acid profile of four strains at the end of 6-day shake flask cultivations at 28 ± 1 °C under nitrogen limiting conditions (C/N = 60). H: H222 wild-type strain, HP: H222 *ΔPOX1-6*, HPDD: H222 *ΔPOX1-6 ΔLEU2 + DGA1 DGA2*, HPDDS: H222 *ΔPOX1-6 ΔLEU2 + DGA1 DGA2 ΔSNF1*.

Figure 7 re-presents the time-course of glycerol consumption and metabolite production for the HPDDS strain. This strain produced citric acid and lipid at a yield of about 0.59 g/g and 0.05 g/g of consumed glycerol, respectively. Therefore, citric acid is considered to be the major bioproduct of this engineered strain. Previous research has reported citric acid production at a yield of 0.93 g/g of glucose hydrol from this yeast [38]. Another study reported a mass yield of 0.90 g citric acid from each gram of glycerol containing waste [34]. Similarly, the lipid titer, content, and yield of 2.82 g/L, 0.39 g/g, and 0.1 g/g of glucose, respectively, were also reported for the *SNF1*-deleted strain overexpressing the *DGA2* gene [36]. That study also reported a synergistic effect between the *SNF 1* deletion and *DGA2* overexpression when combined in the same strain.

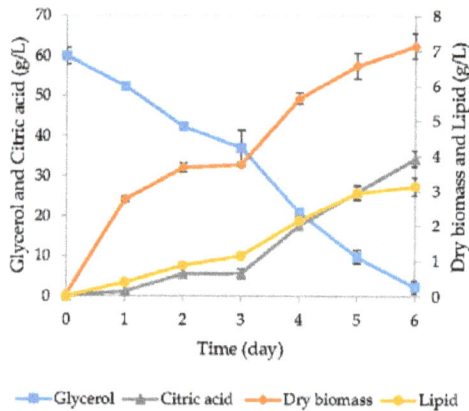

Figure 7. Time course study of glycerol consumption and metabolite production by the HPDDS strain during 6-day shake flask cultivations at 28 ± 1 °C under nitrogen limiting conditions (C/N = 60). Error bars represent standard deviation of *n* = 3. HPDDS: H222 *ΔPOX1-6 ΔLEU2+DGA1 DGA2 ΔSNF1*.

3.2. Fermentation Study

Figure 8 presents data from the time course study of the feedstock consumption and the metabolite production from the best strain (HPDDS) in the 1-L bioreactor. The average of the two batch fermentation rounds is shown.

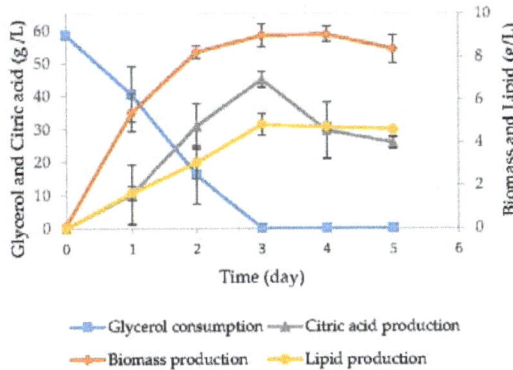

Figure 8. Time course study of glycerol consumption and metabolite production by HPDDS strain during 5-day batch fermentation at 28 °C under nitrogen limiting conditions (C/N = 60). Data from 2 rounds of batch fermentation. HPDDS: H222 ΔPOX1-6 ΔLEU2+DGA1 DGA2 ΔSNF1.

The engineered strain completely utilized 60 g/L glycerol within the three days of fermentation. After that, the strain began to use the secreted citric acid as the carbon source. This observation can explain citric acid reduction during the remaining days of fermentation. The maximum biomass production significantly ($p < 0.05$) increased, from 7.15 g/L in the shake flask to 9 g/L in the bioreactor for the same strain. A slight reduction in biomass was observed on the last day of the fermentation. This may be due to biomass loss during the precipitation of cells containing a relatively large amount of intracellular lipids (over 50%).

Fermentation in the bioreactor ($p < 0.05$) significantly enhanced citric acid and lipid production over the shake flask culture. Citric acid reached the concentration of up to 45 g/L with a yield of 0.75 g/g. The productivity of citric acid was also increased from 0.39 g/L·h to 0.63 g/L·h. Similarly, citric acid was obtained at a productivity of 0.4 g/L·h by growing this yeast in an unbuffered medium with glucose under nitrogen-limiting conditions in flasks [13]. In that study, citric acid was produced at a titer and yield of 35 g/L and 0.43 g/g, respectively, from high initial raw glycerol (80 g/L). In another study, citric acid was obtained at a volumetric productivity of 0.89 to 1.14 g/L·h from *Y. lipolytica* when grown on pure or impure glycerol in a fermenter [34]. In fact, this product was less stable in our bioreactor cultivation over the shake flask fermentation due to a faster glycerol consumption rate and perhaps more control over aeration conditions. Our results show that scale-up of the engineered HPDDS strain in a 1-L bioreactor also enhanced lipid yield and productivity. The peak of lipid production was reached during three days of fermentation at 4.8 g/L, accounting for 53% of DCW, with a productivity of about 0.07 g/L·h. Fontanille et al., used glycerol at a concentration of 80 g/L with a C/N ratio of 62 in batch bioreactor experiments [39]. They reported biomass, lipid titer, and content of 42 g/L, 16 g/L, and 38%, respectively. In another study, crude glycerol with 60% impurity was added to an industrial effluent at a concentration of 4%. Utilization of the feedstock resulted in a lipid titer of 2.21 g/L, which was increased to 2.81 g/L after scale-up [40].

It is not surprising that we observed the constitutive lipid accumulation phenotype simultaneously with biomass propagation. This was due to the inactivation of *SNF1*. The highest biomass, citric acid, and lipid content was reached within the three days of the bioreactor fermentation. This was faster than the shake flask cultivation, in which peaks were reached in six days. Likewise, a constitutive fatty acid production was reported in both growth and oleaginous media after deleting *SNF1*, a negative regulator of lipid accumulation [5]. Another study introduced an approach for constitutive lipid accumulation and citric acid secretion by deleting the *PHD1* gene (YALI0F02497g) encoding the synthesis of 2-methylcitrate dehydratase [37]. The researchers reported a citric acid titer and yield of 57.7 g/L and 0.91 g/g of waste glycerol, respectively, under nitrogen-limiting conditions. However, their maximum lipid content was obtained under nitrogen excess conditions at a titer of 0.98 g/L (accounting for up to 31% of DCW) from 60 g/L waste glycerol. Generally, the *de novo* production of both lipid and citric acid is biochemically equivalent under nitrogen-limiting conditions [37]. The predominance of one occurs at the expense of another depending on the strain. For instance, citric acid was secreted as a major bio-product when the lipid content was less than 22% [13,41].

We also examined the fatty acid composition of accumulated lipid. The results are shown in Figure 9.

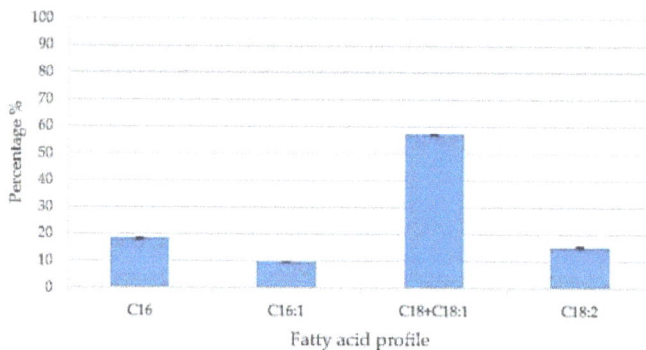

Figure 9. Fatty acid composition of accumulated lipid in HPDDS strain from the bioreactor fermentation. HPDDS: H222 Δ*POX1-6* Δ*LEU2* + *DGA1 DGA2* Δ*SNF1*.

In this study, the major accumulated fatty acids were, in order of abundance, oleic, palmitic, linoleic, stearic, and palmitoleic acid. The ratio of C18:1/C18 was about 8, based on the associated peak heights. Similarly, oleic acid and palmitic acid accounted for 44% and 36% of total fatty acids when glycerol was used in the production medium [40]. Papanikolaou et al. (2013) reported that oleic acid and palmitic acid constitute 52% and 21% of total fatty acids in the engineered strain devoid of *PHD1* gene using glycerol at a concentration of 60 g/L under nitrogen-limiting conditions [37].

3.3. Comparative Study of LEU2 Expression and Leucine Addition

We also examined the effect of *LEU2* expression over leucine supplementation, and the results are listed in Table 4. In order for *LEU2* expression to occur, we transferred the low-copy shuttle vector pGR12 carrying the leucine marker into the HPDDS strain and finally grew the transformants on YNB-Leu (C/N = 60) for six days. We used the shuttle vector-based expression since the *LEU2* locus on the chromosome was used for the site-specific integration of the *DGA 1, 2* genes. For leucine supplementation, we grew the HPDDS uracil$^+$ leucine$^-$ on the YNB-Ura (C/N = 60) medium. This defined medium contained 1.92 g/L of Drop-out Synthetic Mix minus Uracil supplement, with about 20% leucine content.

A higher rate of glycerol consumption was observed for the leucine supplementation treatment over the *LEU2* expression treatment. The level of citric acid and lipid production was also significantly ($p < 0.05$) higher for the former treatment. In the case of leucine supplementation, the lipid content reached 46%, while it was 30% for the *LEU2* expression treatment. This highlights the stimulatory role of leucine in directing carbon flux toward citric acid and lipid production under nitrogen-limiting conditions in the HPDDS strain. We postulate that the concentration of exogenous leucine that was added to the media is higher than those endogenous one, which was synthesized by the *LEU2* carrying low-copy replicative vector. Higher leucine concentration can create a stronger signaling effect for metabolite production in the strain.

Table 4. The effect of leucine supplementation over *LEU2* expression on biomass and metabolite production by the HPDDS strain. HPDDS: H222 Δ*POX1-6* Δ*LEU22+DGA1 DGA2* Δ*SNF1*. a, b Values superscripted with dissimilar letters in same column are significantly different ($p < 0.05$).

Strain and Culture Medium	Residual Glycerol	DCW (g/L)	Citric Acid (g/L)	Lipid (g/L)
HPDDS ura⁺ leu⁻, YNB-Ura	4.93 ± 0.74 [a]	7.75 ± 0.89 [a]	28.36 ± 4.36 [a]	3.6 ± 0.18 [a]
HPDDS ura⁺ pGR12-leucine, YNB-Leu	2.84 ± 0.45 [b]	6.73 ± 0.24 [a]	14.21 ± 1.12 [b]	1.99 ± 0.15 [b]

4. Discussion

The metabolism of glycerol to citric acid [42] and lipogenesis pathway [43] in *Y. lipolytica* has been studied before. Glycerol is assimilated into the cell via facilitated diffusion and is subsequently phosphorylated [41]. *Y. lipolytica* has a unique glycerol metabolism that is dedicated to glycerol-3-phosphate (G3P) and TAG synthesis [44]. Biosynthesis of TAG requires G3P backbone that is acylated by fatty acids [45]. The *de novo* synthesis of fatty acids uses starting units of acetyl-CoA and malonyl-CoA, as well as the cofactor and energy in the form of NADPH and ATP [46]. In *Y. lipolytica*, acetyl-CoA carboxylase (*ACC1*) catalyzes the first committed step of fatty acid synthesis. This involves the conversion of acetyl-CoA to malonyl-CoA precursor. A constant supply of this precursor is required for biosynthesis of fatty acids and other secondary metabolites in yeast [47]. The saturated fatty acids released from fatty acid synthetase complex (FAS), in the form of acyl-CoA [48], are transferred to the endoplasmic reticulum (ER). Fatty acids may undergo further elongation and desaturation before being incorporated into complex lipids though the Kennedy pathway [3]. In the final step of the TAG synthesis pathway, *DGA1*, YALI0E32769g on the lipid droplet (LD) membrane, and *DGA2*, YALI0D07986g in the ER, play prominent roles in acylation of diacylglycerol to produce TAG, which is stored in LDs especially during the stationary phase [49–51]. Therefore, the overexpression of *DGA1* and *DGA2* results in enhanced lipid accumulation [3,52–59]. TAG synthesis and remobilization is a dynamic process. The latter is carried out by the action of lipases and hydrolases [60] in response to the change of cellular or environmental conditions. Deletion of one or two intracellular lipases, *YlTGL3* and *YlTGL4* led to a two-fold increase in the capacity of cell to accumulate lipid [61]. Degradation of released fatty acids take places in the peroxisome via β-oxidation pathway. Intracellular lipid degradation (turn over) can occur along with citric acid secretion in stationary phase [37]. Studies show that nitrogen-limiting conditions also promotes citric acid secretion [62]. In fact, lipid synthesis and citric acid production compete for acetyl-CoA precursors [34].

In our study, glycerol was utilized by all four strains for biomass, lipid, and citric acid synthesis under nitrogen-limiting conditions. The results show that our genetic engineering strategies negatively affected biomass production to different degrees. Inactivation of *SNF1* regulatory gene compensated for part of the biomass loss by increasing the lipid content. The inactivation of *SNF1* gene can enhance the activity of Acc1 and creates larger pool of malonyl-CoA [63]. This generates a push that when combined with the *DGA1* and *DGA2* overexpression, can result in promotion of lipid accumulation accounting for some biomass recovery.

In terms of citric acid production, the citric acid producer H222 wild-type strain showed a citric acid peak of 5 g/L under nitrogen-limiting conditions. Prior study showed that citric and isocitric acids occurs and were excreted by the *Y. lipolytica* control strain at a concentration of 9 g/L [5]. Generally, in wild-type strains, citric acid can also be the product of lipid degradation. However, the availability of glycerol as an external carbon source can lessen the contribution of this phenomenon to citric acid production by reducing internal lipid turnover. In fact, fatty acid remobilization rate decreases in response to a large amount of glycerol [37]. Our engineered strains of HP, HPDD, and HPDDS showed significantly ($p < 0.05$) higher citric acid production compared to the wild strain. However, the exhaustion of glycerol was followed by reduction in citric acid due to its consumption by the cells. Our best-engineered strain, HPDDS, produced citric acid at the titer and yield of more than 45 g/L and 0.75 g/g of pure glycerol in both the flask and bioreactor. The citric acid production at a titer of more than 35 g/L was reported in the literature using a high concentration of industrial glycerol (80 and 120 g/L) [13]. Another study screened tens of strains for citric acid production. They reported a citric acid production range of 71 to 98 g/L by *Y. lipolytica* strains when they were fed with pure or impure glycerol under the nitrogen-limiting conditions [34]. The citric acid production can be further enhanced by adjusting the pH and dissolved oxygen [16]. However, it should be noted that the control over oxygen did not significantly affect citric acid production from *Y. lipolytica* grown on glycerol, as sole source of carbon [64]. Regarding the effect of our genetic engineering on lipid content, inactivating the *POX* genes did not significantly ($p > 0.05$) increase lipid content. This was most likely due to the preference of the strain for utilizing glycerol as carbon source over the intracellular lipids [37]. However, this is a beneficial genetic modification for stable lipid accumulation, particularly after exhaustion of the carbon source [45]. In fact, peroxisomal lipid degradation is still active in cells grown on non-fatty acid feedstock during the stationary phase [65]. In terms of impaired peroxisomal degradation, fatty acids undergo activation by cytosolic YlFaa1p and are re-stored in LD [66]. Disruption of β-oxidation is also a good strategy for creating positive synergism with other complementary modifications to the lipid pathway. Disruption of this pathway is often carried out by deleting *POX* genes [44,45]. Overexpression of diacylglycerolacyltransferases upon integration into the genome generates a pull towards the lipid biosynthesis to accommodate more acyl-CoA pool in the lipid droplets. Thus, further diversion of carbon flux into the lipid droplet was achieved by *DGAs* overexpression. Overexpression of these two diacylglycerolacyltransferases in the HPDD strain that was devoid of active β-oxidation increased lipid accumulation to more than 41% of DCW compared to 17% in the wild-type strain. Similarly, the overexpression of diacylglycerol acyltransferase increased the lipid content from 13% to the range of 39–53% of the DCW in the strain without active *PEX10* [58]. Other studies have attempted *DGA1* or *DGA2* overexpression strategy to enhance lipid content in this yeast [3,54–57]. Overall, overexpression of *DGA* genes increased lipid content while decreasing biomass formation in the HPDD strain due to *LEU2* deletion. Our best-engineered strain, HPDDS, produced lipid at a titer of 3.15 g/L in the shake flask, rising to 4.8 g/L in the batch bioreactor cultivation. A lipid titer of 2.6 to 6.5 g/L was obtained by engineered *Y. lipolytica* from 60 grams of sugar using batch fermentation [51,56]. The coupling of *DGA1,2* overexpression and *SNF1* deletion enabled a maximum lipid content of nearly 44% in the best lipid producing strain, HPDDS. This increased to 53% in the bioreactor. In the same fashion, overexpression of *DGA1* in a ΔSNF2 background led to a 2.7 fold increase in lipid content of *S. cerevisiae* from 11% to 27% [67]. They suggested the strategy using high-copy number plasmid for *DGA1* overexpression and supplementing the media with leucine for enhancing lipid accumulation in the ΔSNF2 disruptant. In the same way, deletion of *SNF1* in combination with *ACC1* overexpression had synergistic effect on enhancing production of fatty acid derivative [68]. Similar synergism was also observed by deletion of the *SNF1* gene and overexpression of *DGA2* in *Y. lipolytica* [36]. The effect of *SNF1* deletion on lipid accumulation in *Y. lipolytica* has also been studied. This deletion resulted in constitutive lipid accumulation phenotype and 2.6-fold higher fatty acid content [5]. This may have been achieved by down-regulating the β-oxidation and enhancing the *ACC1* expression level. Snf1 is a global regulator of cellular energy and contributes to fatty acid regulation at various points. Its deletion can cause overexpression of fatty acid synthases (FAS1 and FAS2) and glyceraldehyde-3-phosphate dehydrogenase (GPD) expression [36]. A protein kinase Snf1 also contributes to other signaling pathways, including amino acid metabolism regulation [69]. Inactivation of Snf1 regulator under nitrogen-limiting conditions can promote citric acid and lipid production as carbon overflow metabolites, while low pH and dissolved oxygen values, and *DGA* genes overexpression can favor lipid production in the resultant phenotype. In fact, pH reduction due to acidification imposes physiologic stress for initiation of lipid accumulation in *Y. lipolytica* [70].

Additionally, longer incubation time may result in reduction of citric acid. This is due to its utilization as carbon source, mainly for cellular maintenance.

In our study, the deletion of *SNF1* gene in the HP strain with deficient β-oxidation did not increase lipid accumulation. Therefore, we did not proceed with that mutant strain. Deletion of *SNF1* in the context of inactive β-oxidation can impose a feedback inhibitory effect of acyl-CoA on Acc1 [36]. It is notable that saturated fatty acids can provoke a feedback inhibitory effect on some biosynthetic enzymes of the fatty acid biosynthesis pathway [71]. Moreover, a study reported that the knockout of *SNF1* results in transcriptional pattern that differed from one on lipid accumulation devaluing the important role of this gene in lipid accumulation [69]. One alternative solution is to use hyperactive mutant Acc1 for higher malonyl-CoA production and consequently higher TAG accumulation, comprised of longer chain fatty acids, without need for *snf1* inactivation [72]. This can contribute to a carbon flux re-direction from citric acid to lipid production. The Snf1 kinase has multiple regulatory roles under different conditions [5], so its preservation can prevent unintended metabolic consequences. Thus, we deleted *SNF1* gene after the integration of the *DGA1* and *DGA2* expression cassette in the HPDD strain to create a synergistic effect to improve metabolite production of TAG.

Overall, the disruption of peroxisomal fatty acid degradation pathway, overexpression of *DGA1* and *DGA2* genes, and inactivation of *SNF1* resulted in the prevention of lipid degradation, up-regulation of TAG biosynthesis pathway, a higher supply of malonyl-CoA and carbon flux towards lipid and citric acid production. Other studies also reported the great performance of *Y. lipolytica* for *de novo* lipid production. An engineered strain overexpressing *ACC1* and *DGA1* produced 28.5 g/L biomass with a lipid content of 61.7% DCW during 5 days of fermentation using 90 g L^{-1} of glucose [54]. This was further enhanced by overexpressing delta-9 stearoyl-CoA desaturase (SCD) to reach biomass, lipid titer, and productivity of 80 g/L, 55 g/L, and 1 g/L·h, respectively, from 150 g/L of glucose [73]. Optimization of the genotype and phenotype resulted in 20 g/L of biomass, 15 g/L lipid, and lipid content of 75% DCW [52]. Combination of evolutionary engineering method with float-based screening resulted in biomass, lipid content and productivity of 45 g/L, 87% and 0.51 g/L·h [74].

We obtained constitutive accumulation of lipid and a high yield of citric acid (0.75 g/g of pure glycerol) from our engineered strain in the bioreactor. The H222 strain used in this study has shown promise for production of organic acids mainly in the form of citric acid and isocitric acid (up to 12%) from different feedstock, including glycerol [75]. We tested some culture media compositions and fermentation conditions for our strains. Fermentation using glycerol under controlled conditions in the bioreactor enhanced lipid productivity from 0.02 to 0.07 g/L·h. In fact, bioreactor cultivation enabled higher citric acid and lipid productivity.

Our genetic engineering strategy noticeably increased the C16 and C18:1 fatty acid contents. Generally, long chain saturated fatty acids are stored in LDs, while short and unsaturated ones are mainly utilized in other anabolic activities [76]. Monounsaturated fatty acyl-CoAs are better substrates for acylation over saturated ones [73]. This may be due to their higher reactivity or toxicity against the host compared with the saturated fatty acids [77]. In fact, TAG synthesis creates buffering capacity to detoxify excess unsaturated fatty acids [77].

We also tested the effect of leucine on biomass and metabolite production at two levels, 380 and 980 mg/L, using YNB-Ura media (data not shown). We did not observe any significant difference ($p > 0.05$) caused by leucine in the foregoing concentration range for the H and HPDDS strains. The stimulatory role of this amino acid was notable in the range of 1–100 mg/L for the leucine-auxotrophic HPDDS strain, and this is due to the compensatory role of this supplementation for the *LEU2* deletion. Previously, it was reported that increasing the leucine supplementation from 100 mg/L to 1600 mg/L enhances lipid accumulation by about six times, and results in genotypic complementation in the leucine⁻ strains without active β-oxidation pathway [52]. It was suggested that this is due to the signaling role of this amino acid and its degradation to acetyl-CoA precursors, which can be subsequently diverted to the lipid biosynthesis pathway. This amino acid plays a regulatory role in lipid metabolism [52]. Studies show that an increased level of leucine may promote lipid metabolism in *Y. lipolytica* through down-regulation of amino acid biosynthesis and deviation of flux from that [69]. A recent study showed a correlation between lipid accumulation and regulation of amino acid biosynthesis [69]. It is plausible that the increased leucine as an important sensor molecule induces transcriptional response [52]. This branch-chain amino acid plays a part in synthesis of fatty acids in adipocytes [78]. The possible role of Snf1, protein kinase, in amino acid metabolism regulation [69] can also explain the interaction of this gene manipulation with leucine metabolism.

The nitrogen-limiting conditions can result in the prevention of biomass proliferation and promotion of lipid accumulation and citric acid secretion to deal with the excess carbon flux. In fact, nitrogen-limiting conditions leads to down-regulation of ribosome structural genes [79]. This also results in degradation of intracellular AMP to release NH4+ ions. Depletion of AMP interferes with the TCA cycle, leading to the accumulation of citric acid in mitochondria and eventually in the cytoplasm [21,37]. Part of citric acid is subsequently broken down by ATP: citrate lyase (ACL) to acetyl-CoA that serves *de novo* synthesis of fatty acid [80]. Although excess availability of carbon can induce lipogenesis in engineered strains [52], it can also trigger citric acid secretion into the growth medium [37]. Thus, it is important to adjust glycerol concentration together with the C/N ratio for optimum biomass and lipid production.

In this study, we achieved constitutive lipid accumulation and citric acid secretion under nitrogen-limiting conditions. Papanikolaou et al., observed similar results by the deleting 2-methylcitrate dehydratase-coding gene. However, they used two different conditions of nitrogen limited and nitrogen excessive conditions for higher citric acid and lipid production, respectively [37].

In summary, our engineered strain shows promise for the simultaneous accumulation of lipid and secretion of citric acid under nitrogen-limiting conditions. The generally recognized as safe (GRAS) status of this yeast justifies the suitability of the major bio-products for delivery to food and pharmaceutical industries. It needs to be pointed out that medium optimization was beyond the scope of our study. Further optimizing the production medium in terms of carbon, nitrogen, and leucine contents is suggested to reach a higher biomass, and subsequently a higher production level. Additionally, the direct use of crude glycerol or pretreated feedstock at larger scale fed-batch fermentation can provide further validation for the performance of the platform developed.

5. Conclusions

In this study, we developed an approach for simultaneous lipid accumulation and citric acid secretion using engineered *Y. lipolytica* in batch fermentation. The combination of deleting the fatty acid degrading pathway, overexpressing key TAG synthesizing genes, and manipulating the lipid regulatory system led to the constitutively accumulation of lipid and secretion of citric acid into the media under the nitrogen-limiting conditions. A relatively high yield of citric acid was achieved along with lipid accumulation from glycerol. This engineered yeast biorefinery platform can be refined and integrated in a fed-batch or continuous system for valorization of a glycerol waste stream into citric acid and lipid.

Supplementary Materials: The following are available online at www.mdpi.com/2311-5637/3/3/34/s1.

Acknowledgments: We would like to thank Washington State University for funding this project, Michael Gatter from the Institute of Microbiology TU Dresden, Germany, for providing strains H222 and H222ΔP. We also greatly appreciate valuable technical advices from Scott E. Baker and Amir H. Ahkami (Pacific Northwest National Laboratory); our colleagues in Washington State University: Jonathan Lomber (Central Analytical Chemistry Laboratory), and Yuxiao Xie (Bioprocessing and Bioproduct Engineering Laboratory).

Author Contributions: Ali Abghari developed the concept, designed the experiments, performed the experiments, analyzed the data, and drafted this paper. Shulin Chen revised the manuscript and approved the final version for publication. All authors read and approved the final manuscript.

Conflicts of Interest: The authors declare no conflict of interest.

References

1. Karamerou, E.E.; Theodoropoulos, C.; Webb, C. A biorefinery approach to microbial oil production from glycerol by rhodotorula glutinis. *Biomass Bioenergy* **2016**, *89*, 113–122. [CrossRef]
2. Kolouchová, I.; Maťátková, O.; Sigler, K.; Masák, J.; Řezanka, T. Lipid accumulation by oleaginous and non-oleaginous yeast strains in nitrogen and phosphate limitation. *Folia Microbiol.* **2016**, *61*, 431–438. [CrossRef] [PubMed]
3. Silverman, A.M.; Qiao, K.; Xu, P.; Stephanopoulos, G. Functional overexpression and characterization of lipogenesis-related genes in the oleaginous yeast *Yarrowia lipolytica*. *Appl. Microbiol. Biotechnol.* **2016**, *100*, 3781–3798. [CrossRef] [PubMed]
4. Bellou, S.; Triantaphyllidou, I.-E.; Mizerakis, P.; Aggelis, G. High lipid accumulation in *Yarrowia lipolytica* cultivated under double limitation of nitrogen and magnesium. *J. Biotechnol.* **2016**, *234*, 116–126. [CrossRef] [PubMed]
5. Seip, J.; Jackson, R.; He, H.; Zhu, Q.; Hong, S.-P. Snf1 is a regulator of lipid accumulation in *Yarrowia lipolytica*. *Appl. Environ. Microbiol.* **2013**, *79*, 7360–7370. [CrossRef] [PubMed]
6. Probst, K.V.; Schulte, L.R.; Durrett, T.P.; Rezac, M.E.; Vadlani, P.V. Oleaginous yeast: A value-added platform for renewable oils. *Crit. Rev. Biotechnol.* **2015**, *36*, 942–955. [CrossRef] [PubMed]
7. Koutinas, A.A.; Vlysidis, A.; Pleissner, D.; Kopsahelis, N.; Garcia, I.L.; Kookos, I.K.; Papanikolaou, S.; Kwan, T.H.; Lin, C.S.K. Valorization of industrial waste and by-product streams via fermentation for the production of chemicals and biopolymers. *Chem. Soc. Rev.* **2014**, *43*, 2587–2627. [CrossRef] [PubMed]
8. Madzak, C. *Yarrowia lipolytica*: Recent achievements in heterologous protein expression and pathway engineering. *Appl. Microbiol. Biotechnol.* **2015**, *99*, 4559–4577. [CrossRef] [PubMed]
9. Abghari, A.; Chen, S. *Yarrowia lipolytica* as oleaginous cell factory platform for the production of fatty acid based biofuel and bioproducts. *Front. Bioenergy Res.* **2014**. [CrossRef]
10. Liu, H.-H.; Ji, X.-J.; Huang, H. Biotechnological applications of *Yarrowia lipolytica*: Past, present and future. *Biotechnol. Adv.* **2015**, *33*, 1522–1546. [CrossRef] [PubMed]
11. Xu, P.; Qiao, K.; Ahn, W.S.; Stephanopoulos, G. Engineering *Yarrowia lipolytica* as a platform for synthesis of drop-in transportation fuels and oleochemicals. *Proc. Natl. Acad. Sci. USA* **2016**, *113*, 10848–10853. [CrossRef] [PubMed]
12. Harzevili, F.D. *Biotechnological Applications of the Yeast Yarrowia lipolytica*; Springer: New York, NY, USA, 2014.
13. Papanikolaou, S.; Muniglia, L.; Chevalot, I.; Aggelis, G.; Marc, I. *Yarrowia lipolytica* as a potential producer of citric acid from raw glycerol. *J. Appl. Microbiol.* **2002**, *92*, 737–744. [CrossRef] [PubMed]
14. Moeller, L.; Strehlitz, B.; Aurich, A.; Zehnsdorf, A.; Bley, T. Optimization of citric acid production from glucose by *Yarrowia lipolytica*. *Eng. Life Sci.* **2007**, *7*, 504–511. [CrossRef]
15. Çelik, G.; Bahriye Uçar, F.; Akpınar, O.; Çorbacı, C. Production of citric and isocitric acid by *Yarrowia lipolytica* strains grown on different carbon sources. *Turkish J. Biochem.* **2014**, *39*, 285–290.
16. Morgunov, I.G.; Kamzolova, S.V.; Lunina, J.N. The citric acid production from raw glycerol by *Yarrowia lipolytica* yeast and its regulation. *Appl. Microbiol. Biotechnol.* **2013**, *97*, 7387–7397. [CrossRef] [PubMed]
17. Pfleger, B.F.; Gossing, M.; Nielsen, J. Metabolic engineering strategies for microbial synthesis of oleochemicals. *Metab. Eng.* **2015**, *29*, 1–11. [CrossRef] [PubMed]
18. Amaral, P.F.F.; Ferreira, T.F.; Fontes, G.C.; Coelho, M.A.Z. Glycerol valorization: New biotechnological routes. *Food Bioprod. Proc.* **2009**, *87*, 179–186. [CrossRef]
19. André, A.; Chatzifragkou, A.; Diamantopoulou, P.; Sarris, D.; Philippoussis, A.; Galiotou-Panayotou, M.; Komaitis, M.; Papanikolaou, S. Biotechnological conversions of bio-diesel-derived crude glycerol by *Yarrowia lipolytica* strains. *Eng. Life Sci.* **2009**, *9*, 468–478. [CrossRef]
20. Rakicka, M.; Lazar, Z.; Dulermo, T.; Fickers, P.; Nicaud, J.M. Lipid production by the oleaginous yeast *Yarrowia lipolytica* using industrial by-products under different culture conditions. *Biotechnol. Biofuels* **2015**, *8*, 1–10. [CrossRef] [PubMed]
21. Kuttiraja, M.; Douha, A.; Valéro, J.R.; Tyagi, R.D. Elucidating the effect of glycerol concentration and C/N ratio on lipid production using *Yarrowia lipolytica* SKY$_7$. *Appl. Biochem. Biotechnol.* **2016**, *180*, 1586–1600. [CrossRef] [PubMed]

22. Dobrowolski, A.; Mituła, P.; Rymowicz, W.; Mirończuk, A.M. Efficient conversion of crude glycerol from various industrial wastes into single cell oil by yeast *Yarrowia lipolytica*. *Bioresour. Technol.* **2016**, *207*, 237–243. [CrossRef] [PubMed]

23. Rywińska, A.; Juszczyk, P.; Wojtatowicz, M.; Robak, M.; Lazar, Z.; Tomaszewska, L.; Rymowicz, W. Glycerol as a promising substrate for *Yarrowia lipolytica* biotechnological applications. *Biomass Bioenergy* **2013**, *48*, 148–166. [CrossRef]

24. Sara, M.; Brar, S.K.; Blais, J.F. Lipid production by *Yarrowia lipolytica* grown on biodiesel-derived crude glycerol: Optimization of growth parameters and their effects on the fermentation efficiency. *RSC Adv.* **2016**, *6*, 90547–90558. [CrossRef]

25. Ledesma-Amaro, R.; Nicaud, J.-M. Metabolic engineering for expanding the substrate range of *Yarrowia lipolytica*. *Trends Biotechnol.* **2016**, *34*, 798–809. [CrossRef] [PubMed]

26. Poli, J.S.; da Silva, M.A.N.; Siqueira, E.P.; Pasa, V.M.; Rosa, C.A.; Valente, P. Microbial lipid produced by *Yarrowia lipolytica* QU21 using industrial waste: A potential feedstock for biodiesel production. *Bioresour. Technol.* **2014**, *161*, 320–326. [CrossRef] [PubMed]

27. Gatter, M.; Förster, A.; Bär, K.; Winter, M.; Otto, C.; Petzsch, P.; Ježková, M.; Bahr, K.; Pfeiffer, M.; Matthäus, F. A newly identified fatty alcohol oxidase gene is mainly responsible for the oxidation of long-chain ω-hydroxy fatty acids in *Yarrowia lipolytica*. *FEMS Yeast Res.* **2014**, *14*, 858–872. [CrossRef] [PubMed]

28. Sambrook, J.; Russell, D.W. *Molecular Cloning: A Laboratory Manual*, 3rd ed.; Coldspring-Harbour Laboratory Press: Cold Spring Harbor, NY, USA, 2001.

29. Gajdoš, P.; Nicaud, J.M.; Čertík, M. Glycerol conversion into a single cell oil by engineered *Yarrowia lipolytica*. *Eng. Life Sci.* **2016**. [CrossRef]

30. Fickers, P.; Le Dall, M.T.; Gaillardin, C.; Thonart, P.; Nicaud, J.M. New disruption cassettes for rapid gene disruption and marker rescue in the yeast *Yarrowia lipolytica*. *J. Microbiol. Methods* **2003**, *55*, 727–737. [CrossRef] [PubMed]

31. Lõoke, M.; Kristjuhan, K.; Kristjuhan, A. Extraction of genomic DNA from yeasts for PCR-based applications. *Biotechniques* **2011**, *50*, 325. [CrossRef] [PubMed]

32. O'fallon, J.; Busboom, J.; Nelson, M.; Gaskins, C. A direct method for fatty acid methyl ester synthesis: Application to wet meat tissues, oils, and feedstuffs. *J. Anim. Sci.* **2007**, *85*, 1511–1521. [CrossRef] [PubMed]

33. Coelho, M.A.Z.; Amaral, P.F.F.; Belo, I. *Yarrowia lipolytica*: An industrial workhorse. *Curr. Res. Technol. Educ. Top. Appl. Microbiol. Microb. Biotechnol.* **2010**, *2*, 930–940.

34. Kamzolova, S.; Anastassiadis, S.; Fatyhkova, A.; Golovchenko, N.; Morgunov, I. Strain and process development for citric acid production from glycerol-containing waste of biodiesel manufacture. *Appl. Microbiol. Biotechnol.* **2010**, 1020–1028.

35. Tchakouteu, S.; Kalantzi, O.; Gardeli, C.; Koutinas, A.; Aggelis, G.; Papanikolaou, S. Lipid production by yeasts growing on biodiesel-derived crude glycerol: Strain selection and impact of substrate concentration on the fermentation efficiency. *J. Appl. Microbiol.* **2015**, *118*, 911–927. [CrossRef] [PubMed]

36. Silverman, A.M. *Metabolic Engineering Strategies for Increasing Lipid Production in Oleaginous Yeast*; Massachusetts Institute of Technology: Cambridge, MA, USA, 2015.

37. Papanikolaou, S.; Beopoulos, A.; Koletti, A.; Thevenieau, F.; Koutinas, A.A.; Nicaud, J.-M.; Aggelis, G. Importance of the methyl-citrate cycle on glycerol metabolism in the yeast *Yarrowia lipolytica*. *J. Biotechnol.* **2013**, *168*, 303–314. [CrossRef] [PubMed]

38. Wojtatowicz, M.; Rymowicz, W.; Kautola, H. Comparison of different strains of the yeast *Yarrowia lipolytica* for citric acid production from glucose hydrol. *Appl. Biochem. Biotechnol.* **1991**, *31*, 165–174. [CrossRef] [PubMed]

39. Fontanille, P.; Kumar, V.; Christophe, G.; Nouaille, R.; Larroche, C. Bioconversion of volatile fatty acids into lipids by the oleaginous yeast *Yarrowia lipolytica*. *Bioresour. Technol.* **2012**, *114*, 443–449. [CrossRef] [PubMed]

40. Louhasakul, Y.; Cheirsilp, B. Industrial waste utilization for low-cost production of raw material oil through microbial fermentation. *Appl. Biochem. Biotechnol.* **2013**, *169*, 110–122. [CrossRef] [PubMed]

41. Makri, A.; Fakas, S.; Aggelis, G. Metabolic activities of biotechnological interest in *Yarrowia lipolytica* grown on glycerol in repeated batch cultures. *Bioresour. Technol.* **2010**, *101*, 2351–2358. [CrossRef] [PubMed]

42. Tomaszewska, L.; Rakicka, M.; Rymowicz, W.; Rywińska, A. A comparative study on glycerol metabolism to erythritol and citric acid in *Yarrowia lipolytica* yeast cells. *FEMS Yeast Res.* **2014**, *14*, 966–976. [CrossRef] [PubMed]

43. Gonçalves, F.; Colen, G.; Takahashi, J. *Yarrowia lipolytica* and its multiple applications in the biotechnological industry. *Sci. World J.* **2014**, *2014*, 14. [CrossRef] [PubMed]
44. Dulermo, T.; Nicaud, J.M. Involvement of the G3P shuttle and β-oxidation pathway in the control of TAG synthesis and lipid accumulation in *Yarrowia lipolytica*. *Metab. Eng.* **2011**, *13*, 482–491. [CrossRef] [PubMed]
45. Beopoulos, A.; Mrozova, Z.; Thevenieau, F.; Le Dall, M.T.; Hapala, I.; Papanikolaou, S.; Chardot, T.; Nicaud, J.M. Control of lipid accumulation in the yeast *Yarrowia lipolytica*. *Appl. Environ. Microbiol.* **2008**, *74*, 7779–7789. [CrossRef] [PubMed]
46. Fakas, S. Lipid biosynthesis in yeasts: A comparison of the lipid biosynthetic pathway between the model non-oleaginous yeast saccharomyces cerevisiae and the model oleaginous yeast *Yarrowia lipolytica*. *Eng. Life Sci.* **2016**. [CrossRef]
47. Tang, X.; Lee, J.; Chen, W.N. Engineering the fatty acid metabolic pathway in saccharomyces cerevisiae for advanced biofuel production. *Metab. Eng. Commun.* **2015**, *2*, 58–66. [CrossRef]
48. Kohlwein, S.D. Triacylglycerol homeostasis: Insights from yeast. *J. Biol. Chem.* **2010**, *285*, 15663–15667. [CrossRef] [PubMed]
49. Beopoulos, A.; Cescut, J.; Haddouche, R.; Uribelarrea, J.L.; Molina-Jouve, C.; Nicaud, J.M. *Yarrowia lipolytica* as a model for bio-oil production. *Prog. Lipid Res.* **2009**, *48*, 375–387. [CrossRef] [PubMed]
50. Zhang, H.; Damude, H.G.; Yadav, N.S. Three diacylglycerol acyltransferases contribute to oil biosynthesis and normal growth in *Yarrowia lipolytica*. *Yeast* **2012**, *29*, 25–38. [CrossRef] [PubMed]
51. Friedlander, J.; Tsakraklides, V.; Kamineni, A.; Greenhagen, E.H.; Consiglio, A.L.; MacEwen, K.; Crabtree, D.V.; Afshar, J.; Nugent, R.L.; Hamilton, M.A. Engineering of a high lipid producing *Yarrowia lipolytica* strain. *Biotechnol. Biofuels* **2016**, *9*, 1. [CrossRef] [PubMed]
52. Blazeck, J.; Hill, A.; Liu, L.; Knight, R.; Miller, J.; Pan, A.; Otoupal, P.; Alper, H.S. Harnessing *Yarrowia lipolytica* lipogenesis to create a platform for lipid and biofuel production. *Nat. Commun.* **2014**, *5*. [CrossRef] [PubMed]
53. Ledesma-Amaro, R.; Nicaud, J.-M. *Yarrowia lipolytica* as a biotechnological chassis to produce usual and unusual fatty acids. *Prog. Lipid Res.* **2016**, *61*, 40–50. [CrossRef] [PubMed]
54. Tai, M.; Stephanopoulos, G. Engineering the push and pull of lipid biosynthesis in oleaginous yeast *Yarrowia lipolytica* for biofuel production. *Metab. Eng.* **2013**, *15*, 1–9. [CrossRef] [PubMed]
55. Beopoulos, A.; Haddouche, R.; Kabran, P.; Dulermo, T.; Chardot, T.; Nicaud, J.-M. Identification and characterization of DGA2, an acyltransferase of the DGAT1 acyl-CoA: Diacylglycerol acyltransferase family in the oleaginous yeast *Yarrowia lipolytica*. New insights into the storage lipid metabolism of oleaginous yeasts. *Appl. Microbiol. Biotechnol.* **2012**, *93*, 1523–1537. [CrossRef] [PubMed]
56. Gajdoš, P.; Nicaud, J.-M.; Rossignol, T.; Čertík, M. Single cell oil production on molasses by *Yarrowia lipolytica* strains overexpressing DGA2 in multicopy. *Appl. Microbiol. Biotechnol.* **2015**, *99*, 8065–8074. [CrossRef] [PubMed]
57. Gajdoš, P.; Ledesma-Amaro, R.; Nicaud, J.-M.; Čertík, M.; Rossignol, T. Overexpression of diacylglycerol acyltransferase in *Yarrowia lipolytica* affects lipid body size, number and distribution. *FEMS Yeast Res.* **2016**, *16*, fow062. [CrossRef] [PubMed]
58. Xue, Z.; Sharpe, P.L.; Hong, S.-P.; Yadav, N.S.; Xie, D.; Short, D.R.; Damude, H.G.; Rupert, R.A.; Seip, J.E.; Wang, J. Production of omega-3 eicosapentaenoic acid by metabolic engineering of *Yarrowia lipolytica*. *Nat. Biotechnol.* **2013**, *31*, 734–740. [CrossRef] [PubMed]
59. Petrie, J.R.; Vanhercke, T.; Shrestha, P.; Liu, Q.; Singh, S.P. Methods of Producing Lipids. U.S. Patent 9,127,288, September 2015.
60. Athenstaedt, K.; Daum, G. The life cycle of neutral lipids: Synthesis, storage and degradation. *Cell. Mol. Life Sci. CMLS* **2006**, *63*, 1355–1369. [CrossRef] [PubMed]
61. Dulermo, T.; Tréton, B.; Beopoulos, A.; Gnankon, A.P.K.; Haddouche, R.; Nicaud, J.-M. Characterization of the two intracellular lipases of *Y. lipolytica* encoded by *TGL3* and *TGL4* genes: New insights into the role of intracellular lipases and lipid body organisation. *Biochim. Biophys. Acta* **2013**, *1831*, 1486–1495. [CrossRef] [PubMed]
62. Papanikolaou, S.; Chatzifragkou, A.; Fakas, S.; Galiotou-Panayotou, M.; Komaitis, M.; Nicaud, J.M.; Aggelis, G. Biosynthesis of lipids and organic acids by *Yarrowia lipolytica* strains cultivated on glucose. *Eur. J. Lipid Sci. Technol.* **2009**, *111*, 1221–1232. [CrossRef]
63. Shi, S.; Chen, Y.; Siewers, V.; Nielsen, J. Improving production of malonyl coenzyme a-derived metabolites by abolishing snf1-dependent regulation of acc1. *MBio* **2014**, *5*, e01130-14. [CrossRef] [PubMed]

64. Sabra, W.; Bommareddy, R.R.; Maheshwari, G.; Papanikolaou, S.; Zeng, A.-P. Substrates and oxygen dependent citric acid production by *Yarrowia lipolytica*: Insights through transcriptome and fluxome analyses. *Microb. Cell Factories* **2017**, *16*, 78. [CrossRef] [PubMed]

65. Desfougères, T.; Haddouche, R.; Fudalej, F.; Neuvéglise, C.; Nicaud, J.-M. Soa genes encode proteins controlling lipase expression in response to triacylglycerol utilization in the yeast *Yarrowia lipolytica*. *FEMS Yeast Res.* **2009**, *10*, 93–103. [CrossRef] [PubMed]

66. Dulermo, R.; Gamboa-Meléndez, H.; Ledesma-Amaro, R.; Thévenieau, F.; Nicaud, J.-M. Unraveling fatty acid transport and activation mechanisms in *Yarrowia lipolytica*. *Biochim. Biophys. Acta* **2015**, *1851*, 1202–1217. [CrossRef] [PubMed]

67. Kamisaka, Y.; Tomita, N.; Kimura, K.; Kainou, K.; Uemura, H. DGA1 (diacylglycerol acyltransferase gene) overexpression and leucine biosynthesis significantly increase lipid accumulation in the Δsnf2 disruptant of *saccharomyces cerevisiae*. *Biochem. J.* **2007**, *408*, 61–68. [CrossRef] [PubMed]

68. Feng, X.; Lian, J.; Zhao, H. Metabolic engineering of saccharomyces cerevisiae to improve 1-hexadecanol production. *Metab. Eng.* **2015**, *27*, 10–19. [CrossRef] [PubMed]

69. Kerkhoven, E.J.; Pomraning, K.R.; Baker, S.E.; Nielsen, J. Regulation of amino-acid metabolism controls flux to lipid accumulation in *Yarrowia lipolytica*. *Syst. Biol. Appl.* **2016**, *2*, 16005. [CrossRef]

70. Nambou, K.; Zhao, C.; Wei, L.; Chen, J.; Imanaka, T.; Hua, Q. Designing of a "cheap to run" fermentation platform for an enhanced production of single cell oil from *Yarrowia lipolytica* DSM3286 as a potential feedstock for biodiesel. *Bioresour. Technol.* **2014**, *173*, 324–333. [CrossRef] [PubMed]

71. Liao, J.C.; Mi, L.; Pontrelli, S.; Luo, S. Fuelling the future: Microbial engineering for the production of sustainable biofuels. *Nat. Rev. Microbiol.* **2016**, *14*, 288–304. [CrossRef] [PubMed]

72. Hofbauer, H.F.; Schopf, F.H.; Schleifer, H.; Knittelfelder, O.L.; Pieber, B.; Rechberger, G.N.; Wolinski, H.; Gaspar, M.L.; Kappe, C.O.; Stadlmann, J. Regulation of gene expression through a transcriptional repressor that senses acyl-chain length in membrane phospholipids. *Dev. Cell* **2014**, *29*, 729–739. [CrossRef] [PubMed]

73. Qiao, K.; Abidi, S.H.I.; Liu, H.; Zhang, H.; Chakraborty, S.; Watson, N.; Ajikumar, P.K.; Stephanopoulos, G. Engineering lipid overproduction in the oleaginous yeast *Yarrowia lipolytica*. *Metab. Eng.* **2015**, *29*, 56–65. [CrossRef] [PubMed]

74. Liu, L.; Pan, A.; Spofford, C.; Zhou, N.; Alper, H.S. An evolutionary metabolic engineering approach for enhancing lipogenesis in *Yarrowia lipolytica*. *Metab. Eng.* **2015**, *29*, 36–45. [CrossRef] [PubMed]

75. Egermeier, M.; Russmayer, H.; Sauer, M.; Marx, H. Metabolic flexibility of *Yarrowia lipolytica* growing on glycerol. *Front. Microbiol.* **2017**, *8*, 49. [CrossRef] [PubMed]

76. Papanikolaou, S.; Aggelis, G. Selective uptake of fatty acids by the yeast *Yarrowia lipolytica*. *Eur. J. Lipid Sci. Technol.* **2003**, *105*, 651–655. [CrossRef]

77. Petschnigg, J.; Wolinski, H.; Kolb, D.; Zellnig, G.; Kurat, C.F.; Natter, K.; Kohlwein, S.D. Good fat, essential cellular requirements for triacylglycerol synthesis to maintain membrane homeostasis in yeast. *J. Biol. Chem.* **2009**, *284*, 30981–30993. [CrossRef] [PubMed]

78. Crown, S.B.; Marze, N.; Antoniewicz, M.R. Catabolism of branched chain amino acids contributes significantly to synthesis of odd-chain and even-chain fatty acids in 3T3-L1 adipocytes. *PLoS ONE* **2015**, *10*, e0145850. [CrossRef] [PubMed]

79. Pomraning, K.R.; Kim, Y.-M.; Nicora, C.D.; Chu, R.K.; Bredeweg, E.L.; Purvine, S.O.; Hu, D.; Metz, T.O.; Baker, S.E. Multi-omics analysis reveals regulators of the response to nitrogen limitation in *Yarrowia lipolytica*. *BMC Genomics* **2016**, *17*, 138. [CrossRef] [PubMed]

80. Zhang, H.; Wu, C.; Wu, Q.; Dai, J.; Song, Y. Metabolic flux analysis of lipid biosynthesis in the yeast *Yarrowia lipolytica* using ^{13}C-Labled glucose and gas chromatography-mass spectrometry. *PLoS ONE* **2016**, *11*, e0159187. [CrossRef] [PubMed]

fermentation

MDPI

Article

Codigestion of Untreated and Treated Sewage Sludge with the Organic Fraction of Municipal Solid Wastes

Khalideh Al bkoor Alrawashdeh [1,*,†], Annarita Pugliese [2,†], Katarzyna Slopiecka [2,†],
Valentina Pistolesi [2,†], Sara Massoli [2,†], Pietro Bartocci [2], Gianni Bidini [3] and Francesco Fantozzi [3]

[1]   Mechanical Engineering Department, Al-Huson University College, Al-Balqa' Applied University,
      P.O.Box 50, Al-Huson, 19117 Irbid, Jordan
[2]   Biomass Research Center, University of Perugia, CRB, Via G. Duranti, 06-125 Perugia, Italy;
      pugliese@crbnet.it (A.P.); kathiss@o2.pl (K.S.); pistolesi@crbnet.it (V.P.); massoli@crbnet.it (S.M.);
      bartocci@crbnet.it (P.B.)
[3]   Department of Engineering, University of Perugia, Via Duranti 67, 06-125 Perugia, Italy;
      gianni.bidini@unipg.it (G.B.); francesco.fantozzi@unipg.it (F.F.)
*   Correspondence: khalideh19@yahoo.com or khalideh.alrawashdeh@bau.edu.jo; Tel.: +962-798-053-087
†   These authors contributed equally to this work.

Received: 5 June 2017 ; Accepted: 15 July 2017 ; Published: 27 July 2017

Abstract: Disposal of biodegradable waste has become a stringent waste management and environmental issue. As a result, anaerobic digestion has become one of the best alternative technology to treat the organic fraction of municipal solid wastes and can be an important source of bioenergy. This study focuses on the evaluation of biogas and methane yields from the digestion and co-digestion of mixtures of waste untreated sludge and the organic fraction of municipal solid wastes. These are compared with the results obtained from the digestion and codigestion of mixtures containing waste active sludge and the organic fraction of municipal solid wastes. The two types of substrates were used to perform biomethanation potential tests, in mesophilic conditions (35 °C) at lab scale. It was observed a maximum biogas yield for 100% of untreated sewage sludge, corresponding to 0.644 Nm^3/kg VS and 0.499 Nm^3/kg VS of biogas and methane production respectively. The study also demonstrates the possibility of increasing biogas production up to 36% and methane content up to 94% using waste untreated sludge substrate in both digestion and codigestion, compared to waste active sludge substrate.

Keywords: co-digestion; sewage sludge; methane production; BMP; municipal solid waste

1. Introduction

The huge amount of sewage sludge and of Organic Fraction of Municipal Solid Waste (OFMSW), which are disposed of daily through incineration or land filling constitutes a huge environmental challenge. The European Union regulations demand that biodegradable municipal waste to landfill sites must be reduced by 25% with respect to 1995 levels by 2010 with a further reduction of 65% by 2016, see [1,2]. According to recent estimates of the European Commission, about 88 Mt of bio-waste extracted from municipal solid waste [3] and 10 Mt of Waste of Active Sludge (WAS) dry matter [4,5] are produced annually in the EU-27. However, given the organic content and chemical composition of WAS and OFMSW, they are easily biodegradable in anaerobic conditions that favor decomposition and mineralization producing biogas and residues, that can be used as nutrient soil replacement. Organic waste management through Anaerobic Digestion (AD) represents a useful solution to decrease the environmental impact caused by landfill disposal.

Waste sludge considers treated sludge in three forms: primary sludge, secondary sludge (which is called WAS) and mixture of primary and secondary sludge (thickened sludge). Primary sludge, according to [1], is more easily degradable in anaerobic conditions than WAS. A typical aerobic treatment for sewage is usually performed in a wastewater treatment plant (see Figure 1), with various scales of aerobic duration and sedimentation, in order to reduce the Biochemical Oxygen Demand (BOD) and Chemical Oxygen Demand (COD) of the waste prior to its landfilling or conveying to surface water. However, primary and secondary treatment process releases a significant amount of methane, which is lost to the atmosphere, increasing the environmental impacts and losing potential energy of the sludge [6]. An AD process integrated with the aerobic treatment, would recover a significant amount of biogas for energy production treatment. Moreover, the possibility of treating together OFMSW and sewage sludge to produce biogas in a system eventually integrated with the aerobic treatment has an interesting potential. However the best mixture between OFMSW and sewage, in terms of biogas production, is still under analysis. Also whether to use WAS (secondary sludge) or Primary Sludge (PS) as even untreated sludge is not yet ascertained. In order to evaluate the quality of biowaste to serve as a substrate in anaerobic digestion several methods are used, such as; Anaerobic Biomethanation Potential (ABP), pilot plant and full scale plant test (see Table 1). Various studies are available on anaerobic co-digestion of treated sludge WAS, Primary Sludge (PS), thickened sludge with OFMSW or biowastes. Various studies are available on anaerobic co-digestion of treated sludge WAS, Primary Sludge (PS), thickened sludge with OFMSW or biowastes, see Kolbl [7,8]; however, no one addresses the co-digestion with Untreated Sewage Sludge (USS—fresh sludge without primary and aerobic treatment). Lab scale results show that the co-digestion of WAS-OFMSW could be the most effective way to improve digester performance, according to [2,9–12]. Murto et al. 2004 [2] observed that co-digestion with the high buffered system leaded to imbalance the process; Cabbai et al. [9] indicated that the high acid load of co-digestion substrates leads to the inhibition of AD process; Cavinato et al. [11] reported that thermophilic conditions perform better in a co-digestion process for sludge/biowaste in the terms of biogas production. Kim et al. [13] investigated the effect of different variables (temperature and mixing ratio), reporting that the addition of food waste to WAS digestion increases methane yield. Gomez et al. [14] observed that co-digestion (WAS-OFMSW) with pH and mixing ratio control achieved high methane production. Sosnowski et al. [15] compared the effects of different mixing ratios and determined that high organic load improves biogas yield. Cabbai et al. [9] concluded that certain types of organic waste source (household and supermarkets wastes) in co-digestion with WAS increase methane yield by 47% with respect WAS digestion alone. From all the above reported studies co-digestion of sewage sludge and organic solid wastes appears to be an important strategy the management of urban wastes.

Researchers have also focused on co-digestion technology based on synergisms/antagonisms between substrates [16], showing that a higher concentration of micronutrients in sludge compensates the shortage of OFMSW in a pH environment suitable for AD bacteria. It was shown that aerobic digestion became more stable with the C/N ratio of co-substrates remaining within the desired range of 22–30, see [17]. The co-digestion of WAS-OFMSW at different mixture ratios, was successfully experimented by [9,11,16]; when Solids Retention Time (SRT) is longer than Hydraulic Retention Time (HRT), Volatile Solids (VS) and microbial biomass should be retained for a higher biogas production, hence mixing can be reduced [18]; when co-digesting proteins-rich substrate WAS can provide the required buffering capacity [19]. Improvement in methane yield through co-digestion was achieved with increasing amount of organic waste in wastewater sludge digestion [9,15]. Nevertheless, to improve methane yield there is a limit to the addition of organic matter depending on the digestion conditions and stability (see Table 1). During the AD of a co-substrate microorganisms utilize carbon from 25 to 30 times faster than nitrogen [20] and the nitrogen content in WAS compensates a possible lack of nutrients in OFMSW while their content of lipids increases biogas yield [21]. On the other hand a lipid-rich substrate leads to an increase in Long Chain Fatty Acids (LCFAs) which may form a hydrophobic layer that destabilizes the digestion process [22], affecting bacteria transport and reducing contact between the substrate and the encapsulated bacteria. LCFAs entrapment causes the flotation and inhibition of methanogenic bacteria leading to cellular membrane damage [23]. Hence, the major benefit of WAS and OFMSW co-digestion is to reduce the toxicity within the media [2,15], and to provide other nutrients which are not present at sufficient levels in OFMSW. However, other authors showed negative results in their research, probably attributable to specific characteristics of the digested substrates [2,24]. This research activity provides information on biogas and methane yield of Untreated Sludge (USL) in different conditions and mixtures with OFMSW, given the lack of experimental data in the Literature, and compares its performance with WAS behavior. Although many studies have analyzed the co-digestion of WAS or thickened sludge with OFMSW still no data is available on the co-digestion of WUS and OFMSW. This works contributes by determining the BMP of different mixture in different condition of WUS-OFMSW and WAS-OFMSW in batch reactors, the sludge feedstock was taken from a sewage treatment plant in central Italy, while OFMSW consisted of household organic waste, collected in Perugia municipality area.

Table 1. Biogas and methane yields in similar studies.

Composition	SPG	CH$_4$	Method	References
(100% WAS)	0.390 (m^3/kg VS)	64%	BMP	[9]
(41.5% WAS-58.5% OFMSW)	0.620 (m^3/kg VS)	n.r.	BMP	[9]
(50% WAS-50% OFMSW)	0.34 (m^3/kg VS)	60%	Pilot scale	[11]
(100% WAS)	0.15 (m^3/kgVS)	61.8%	Pilot scale	[11]
(50% WAS-50% OFMSW)	0.35 (m^3/kg VS)	60%	Full Scale	[11]
(75% WAS-25% OFMSW)	0.45 (m^3/kg VS)	53.8%	Pilot scale	[9]
(41% WAS-59% OFMSW)	0.43 (m^3/kg VS)	64%	Full scale	[25]
(77% TAS-23% (KW & FWP))	0.38 (Nm^3/kg VS)	n.r.	Pilot scale	[26]
((60% PS & 40% WAS)-OFMSW)	0.6 (m^3/kg VS)	n.r.	Full scale	[27]
(100% biological sludge)	0.27 (m^3/kg VS)	60%	BMP	[12]
(80% OFMSW-20% biological sludge)	0.22 (m^3/kg VS)	n.r.	BMP	[27]

2. Materials and Methods

2.1. Sewage Sludge and Organic Fraction of Municipal Solid Wastes Samples

Both WUS and WAS were obtained from a wastewater treatment plant which is part of network that serves a 150,000 citizens of the town of Perugia in central Italy; the plant collects the households effluents (a population equivalent of 30,000) and the industrial wastewater of the area. With reference to Figure 1, the plant layout consists of a series of treatment units (primary clarifier and secondary clarifier) and a final drying section for the sludge generated to be used in composting.

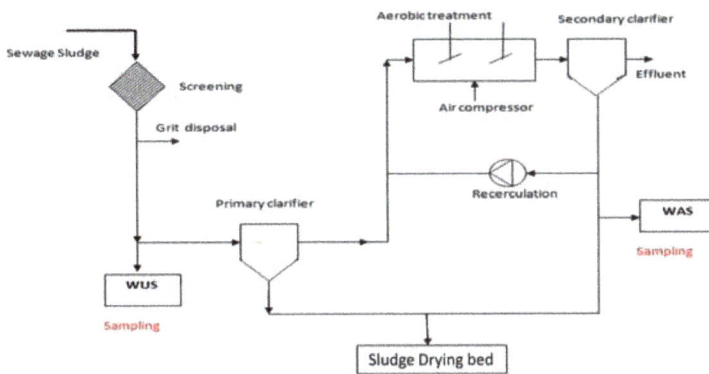

Figure 1. WWTP of the Umbrian Water management company.

As received fresh sewage was collected on site before the primary and secondary treatment unit during a normal working day with constant time steps (5 L sludge/10 min) summing up to 100 L. WUS was stored for 24 h. While WAS was collected from the exhaust of the secondary unit. The samples used in the BMP tests were prepared according to UNI 5667-13/2000. OFMSW was assembled from households waste with the following concentration in weight: 45% vegetables and fruit waste, 35% residuals of bread and pasta, 10% rice, 5% paper and 5% coffee; the materials were eventually homogenized with an electric mixer. The samples of OFMSW were obtained according to CENT/TS 14778-1, CENT/TS 14779. The inoculum was collected from a nearby anaerobic digestion plant.

2.2. Samples Analysis

The chemical and physical analyses were performed according to standard methods by means of a thermobalance TGA 701 LECO and a Truspec CHN LECO [28]. Moisture, ash and volatile solids content were obtained according to CEN/TS 14774; CEN/TS 14775 and CEN/TS 15148. To perform proximate analysis the samples were heated according to CEN/TS14780 and the ultimate analysis was carried out according to CEN/TS 15104. pH of substrates was measured continuously throughout the tests with a probe (Hanna Instruments HI 9124, double junction electrode, resolution 0.01). WUS, WAS, OFMSW and inoculum used in the test had the characteristics shown in Table 2.

Table 2. Characteristics of the substrates.

	Moisture (%)	Total Solids (%)	Volatile Solids (%)	Ash (%wb)	Fixed Carbon (%wb)	pH	C/N
WUS	93.97	6.03	4.33	1.7	0	7.01	12
WAS	95	5.0	3.12	1.88	0	7.3	8.6
OFMSW	76.22	23.78	19	1.9	2.88	6.25	35.8
Inoculum	97.0	3	2.06	0.94	0	7.78	13

2.3. Experimental Setup

Biomethane Potential (BMP) tests were carried by means of in house designed vessels, with a global capacity of about 1 L realized in Boro-silicate glass and equipped with a major neck connected to a pressure sensor UNIK 5000 GE Measurement & Control. In addition, minor necks with plugs to guarantee sealage during the test are used to get biogas samples and to measure pH. Biogas production is continuously derived from pressure (UNIK 5000, accuracy to 0.04% and stability typically 0.05%), and recorded for post processing. Biogas was sampled and analyzed by a gaschromatograph (490 micro GC, Agilent Technologies, Santa Clara, CA, USA), Helium and Argon were used as a carrier gas with a flow rate of 10 mL/min. Temperature of detector injector and columns were 180 °C, 100 °C, and 80 °C respectively. Biogas in excess was continuously vented to avoid pressurized conditions and explosion risks. For a throughout description of the laboratory equipment see [28–31].

2.4. Experimental Procedure

The vessels were filled up to 20% of their volume with different mixtures of WUS-OFMSW and WAS-OFMSW prepared similarly to the WAS-OFMSW mixture tested in [9,11,12,25–27].

The ratio concentrations of WUS to OFMSW and WAS to OFMSW by weight were: 50:50, 70:30 and 100% in weight respectively. The vessels were then tightly closed and flushed with N_2 to vent the air and remove O_2. Then sensors are applied and the vessels are sealed and immersed in a thermostatic bath (see Figure 2) in mesophilic conditions (approximately 35 ± 0.5 °C).

All the vessels were shaken manually two times a day (for 1 min) during the initial two weeks of the test period as recommended by Reference [18,32]. In order to avoid formation of the buffer layer within the substrate, and to insure that substrate molecules and bacterial can join, see [32]. Table 3 shows the concentration of the vessels.

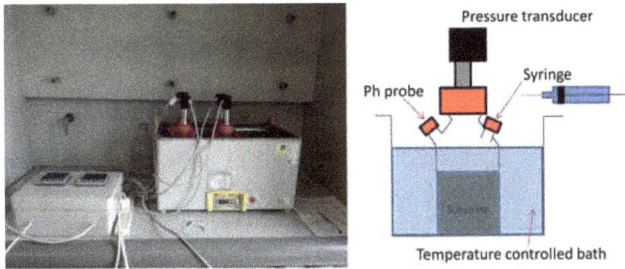

Figure 2. Laboratory equipment used in BMP tests.

Table 3. Substrates characteristics and compositions of all the vessels.

Vessels	Substrate	Moisture (%)	VS/TS	C/N	pH
1 & 2 (70%-30% weight) WUS:OFMSW	107 g WUS 46 g OFMSW 47 g Inoculum	90.6	83.3	17.67	6.9 (vessel 1) 7.0 (vessel 2)
3 & 4 (50%-50% weight) WUS:OFMSW	59 g WUS 59 g OFMSW 82 g Inoculum	90.0	85.6	19.37	6.4 (vessel 3) 6.1 (vessel 4)
5 (100% weight)	200 g Inoculum	97	68.7	13	7.8
6 (100% weight)	200 g WUS	93.97	71.8	12	7.0
1* & 2* (70%-30%) WAS:OFMSW	107 g WAS 46 g OFMSW 47 g Inoculum	91.15	81.2	15.85	6.9 (vessel 1) 7.0 (vessel 2)
3* & 4* (50%-50% weight) WAS:OFMSW	59 g WAS 59 g OFMSW 82 g Inoculum	90.28	84.56	18.37	6.9 (vessel 1) 7.0 (vessel 2)
6* (100% weight)	200 g WAS	95	62.4	8.6	7.3

Symbol "*" denotes the vessels containing WAS.

Measurements of pH were performed with a probe on the substrates every day during the initial four weeks, then performed once a week due to the relative stability pH value. During the initial phase pH value was corrected every three days by adding 1.0 mL of KOH to vessels 1 & 2, 1.3 mL of KOH to vessels 3 & 4. While the vessels which contain WAS substrate (vessels: 1*, 2*, 3*, 4* & 6*) and mono-substrate of WUS (vessel 6) did not need correction.

3. Results and Discussion

Daily and cumulative production of biogas of all vessels are presented respectively in Figure 3 and the specific of biogas and methane production in Figure 4. The duration of the test was around a hundred days. During start-up vessels 1, 2, 3 & 4 (co-mixture of WUS-OFMSW) pH decreased significantly reaching high acid values; pH control increased the production rate of the vessels 3 & 4, while production remained low for vessel 1 & 2. Total and volatile solids were determined, using a syringe and maintaining anaerobic conditions, every 20 days to track organic matter decomposition. Decomposition rates vary among WUS and WAS substrates of both co-digestion and mono-digestion. The highest rates of VS and TS were observed with WUS substrates in both types of digestion as shown in Figure 5.

Daily production of vessels 1 & 2 (70/30 of WUS-OFMSW) and 1* & 2* (70/30 of WAS-OFMSW) show similar trends. An initial phase with a low production rate is followed by a high production rate phase until a plateau is reached and rustained in the final decaying phase. However, the performance and biodegradation of substrate of 70/30 of WUS-OFMSW (vess. 1 & 2) is higher than the one of the 70/30 of WAS-OFMSW mixture of (vess. 1* & 2*). According to the biodegradation behavior of vessels of 1 & 2 (Figure 5) a slower acidogenic phase is present with respect to vessels 1* & 2*, a similar behavior was described by [25]. This affected methane production, which for vessels of 1 & 2 was measured by µGC in the range of 40–44% volume of biogas. Overall, biogas and methane production of vessels 1 & 2 are significantly higher than those produced by vess. 1* & 2* (see Figure 4 and Table 4). This test was carried out with a high buffering capacity and a balanced process for vessels 1* & 2* in agreement with [33].

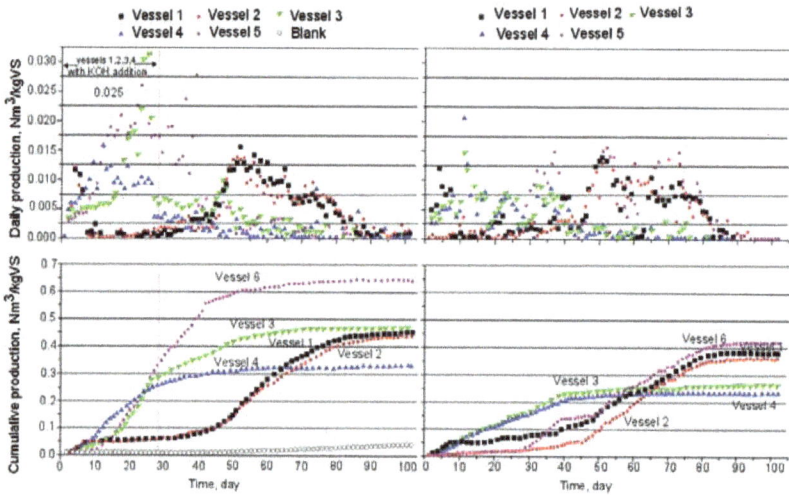

Figure 3. Daily and cumulative performance of biogas production of WAS and WUS digestion in both co-digestion and mono-digestion.

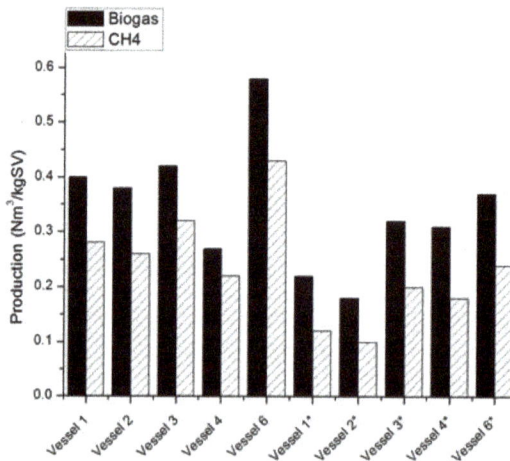

Figure 4. Specific biogas and methane production of all vessels.

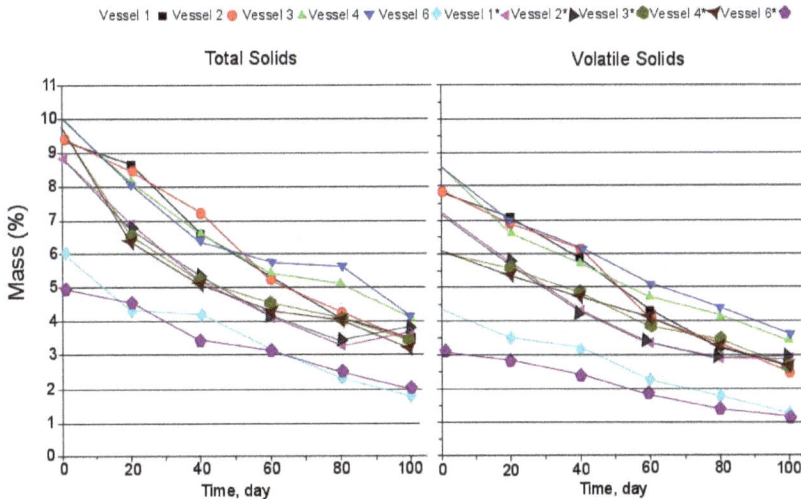

Figure 5. Total and volatile solids decomposition determined every 20 days.

Vessels 3* & 4* were characterized by slower degradable substrate compared to vessels 3 & 4 (even their 1st degradation phase started on the first day of the test). Moreover, the daily yield of vessel 4 indicated that the biogas production rate increased sharply at the beginning and then gradually decreased with constant and low rate for the last day of the test. While daily production of vessel 3 was starting to climb steeply and then drop sharply after that it became progressively stable (after day 25 till 70). Daily production of vessel 3* & 4* it shows a repeated pattern of degradation that occurs every 15 days during the test till day 45. However the vessels 3* & 4* have lower methane content than Vessels 3 & 4 (see Figure 4). In the case of vessel 4, it has the lowest yield of biogas and by tracing the degradation behavior of it, that seemed to descend back down at day 50 (see Figure 3). Moreover, in the initial phase of the test all the conditions of the process were normal, especially the high concentration of OFMSW (high acidic load) which required a correction of pH with 1.3 mL KOH and the 2nd phase was carried out with exhaustion of the buffering capacity (the carbonate system at pH value closed to 8) during the last period, a similar trend was reported by [34]. The test confirms that a neutralized substrate can be controlled (vess. 3 & 4), and that despite the acidic environment of the substrate with higher content of organic load (mixtures WUS-OFMSW) but they released production significantly. Anyway, the average yields of biogas and methane of co-mixture 50/50 of both WAS and WUS to OFMSW are lower than yields of all other mixtures and mono-digestion, due to the solid retention time which increased under the high solid concentration, that as reported by Bolzonella et al. [25], the SGP decreased from 0.18 to 0.07 m^3/kg VS fed when increasing the solid retention time in AD process. Both mono digestion process of WAS and WUS (vessels 6 and 6*) were stable and balanced with high biogas production. In fact, both substrates are characterized by lipids, so that both required a long retention time due to slow biodegradation, a similar behavior (sludge digestion) was described by [9,35]. Vessel 6* required sufficient time to reach the phase of biogas generation, vessel 6 produced more because of a higher organic load than vessel 6*. Vessel 6 achieved the highest peak of methane contents (0.035 Nm^3/kg VS) in day 40 and was more than 76.6% of CH_4 (see Figure 3). The test confirmed that significant differences of biogas yield and methane content from WUS compared to yields of WAS. However the statistical analysis of specific production of biogas and methane confirmed that the WUS and co-digestion mixture (WUS-OFMSW) yielded higher than those produced by pure sewage sludge and by the co-digestion mixture (WAS-OFMSW), as shown in Table 4. these data should be further scaled up to pilot pants scale and to industrial scale based also on the influence of Organic Loading Rate (OLR) and Hydraulic Retention Time (HRT).

Table 4. Biogas and methane yield and VS removal percentage of co-mixtures and mono-substrates, (with variance 0.00311 for biogas and 0.00273 for methane).

	Biogas (Nm³/kg VS)	Methane (Nm³/kg VS)	VS Removed
(WUS:OFMSW)			
(70%-30% weight)	0.444	0.331	66
(50%-50% weight)	0.399	0.315	60
(100% WUS)	0.644	0.499	72
(WAS:OFMSW)			
(70%-30% weight)	0.370	0.243	58
(50%-50% weight)	0.245	0.162	58
(100% WUS)	0.410	0.283	65

Biogas and methane SPGs of the blank (inoculum) are 0.083 Nm^3/kg VS and 0.024 Nm^3/kg VS respectively. These were subtracted from the production of the mixtures, based on the different masses of inoculum which were present. Figure 6 illustrates the linear fit of biogas and methane production of WUS-OFMSW and WAS-OFMSW.

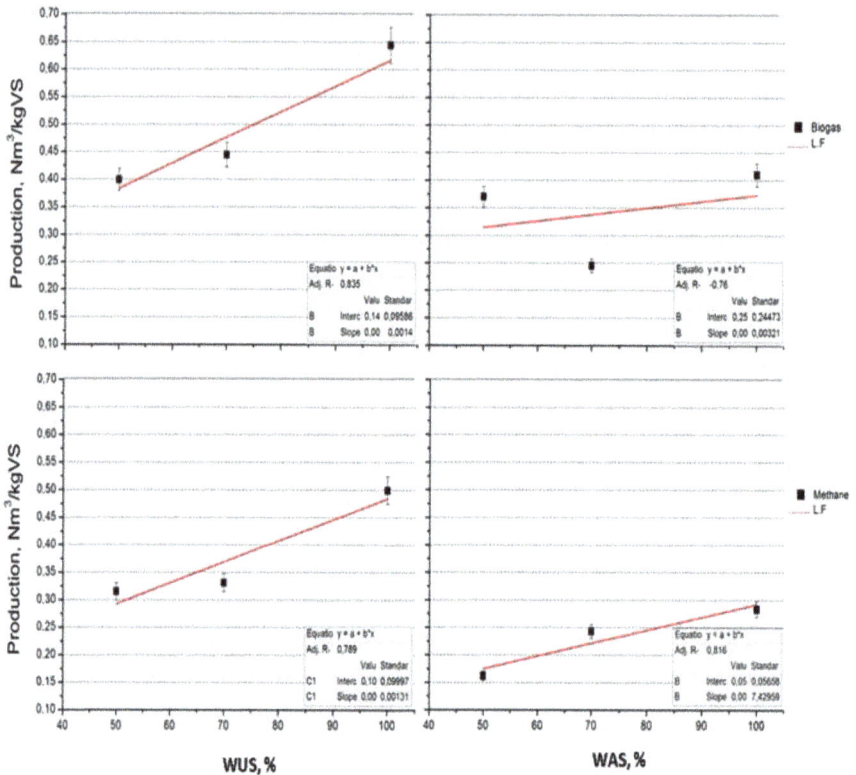

Figure 6. The relationship of WAS and WUS concentration vs biogas and methane production.

With the increase of the amount of sludge (WUS and WAS) it can be observed a significant growth in methane and biogas production, that's consistent with what mentioned in the Literature [9,15,25]. But in fact, the amount of the production growth from WUS digestion is a higher than that of WAS digestion, that explains the role of the WUS which have higher amount of VS than WAS and its role to improve the C/N ratio of the substrate. Causing ideal condition to increase the production of biogas and methane. Substrates of 50/50 have a higher hydraulic potential compared to substrates of 70/30, where the addition of OFMSW improve and accelerate the hydrolysis of WUS digestion (Figure 5), that which agrees to [36]. Vessels 1, 2, 3, 4 and 6 had a higher content of the initial organic load with respect to vessels 1*, 2*, 3*, 4* and 6* respectively, all the vessels contain WUS were characterized by a richer lipid than vessels contain WAS substrate that required a longer time for degradation. In the case of mono digestion and codigestion of WAS and OFMSW, pure WAS (vess. 6*) achieved a higher methane content than co-digestion, that agrees with [12]. The experimental results indicate that USS digestion improved the biogas yield up to 20%, 57% and 62% of 70/30 (WUS-OFMSW), 50/50 (WAS-OFMSW) and 100% of USS respect to the same ratio of WAS-OFMSW and 100% of WAS, CH_4 yield was increased up to 94% by (50/50) WUS-OFMSW. Moreover, our results which were obtained for WAS and its mixtures with OFMSW accord with results reported in Reference [9,11] in the terms of behavior and biogas/methane yield, and CH_4 yields were superior to those obtained in Reference [15,37–39]. The co-digestion between WUS and OFMSW is a suitable solution for waste management and an alternative renewable energy source from the conversion of wastes into biogas, these observation accords with [40]. However, the results indicated that the digestion of pure WUS (Table 4 and Figure 4) is the best substrate for anaerobic digestion, where it achieved the highest biogas and methane yield: 0.644 Nm³/kg VS and 0.499 Nm³/kg VS respectively. Additionally results indicated that a higher sewage sludge content could significantly increase biogas and methane production rate.

4. Conclusions

The experiment has shown that an 100% of WUS mono-digestion is the optimal substrate for biogas and methane production. As expected using WUS substrates, positively affects biogas and methane production, with a significant increases compared to the use of treated sewage sludge extracted after aerobic treatment, and increases biogas yield in the range of 20% to 62%, methane content in the range of 36% to 94%, and an increase in the range of 3 to 10% in removable VS of WUS-OFMSW co-digestion compared to WAS-OFMSW substrate. Biogas and methane cumulative production during co-digestion of WUS-OFMSW increases notably when increasing the WUS to OFMSW ratio. The influence of seasonal change in the characteristics of the sludge will be taken into account in future work.

Acknowledgments: The authors gratefully acknowledge the contribution of Pippi of Umbbria Acque Spa for providing the WUS and WAS, and Alessandro Iraci and Giacomo Iraci of Agricola IRACI BORGIA s.s. for providing the inoculum.

Author Contributions: The work was entirely developed and managed by Khalideh Al bkoor Alrawashdeh under her PhD project at the Biomass Research Centre, Perugia Italy. The superivors Francesco Fantozzi and Gianni Bidini have checked the work. Other colleagues have helped performing experimental campaign, such as Annarita Pugliese, Katarzyna Slopiecka, Valentina Pistolesi, Sara Massoli and Pietro Bartocci.

Conflicts of Interest: The authors declare no conflict of interest.

Abbreviations

The following abbreviations are used in this manuscript:

AD	Anaerobic Digestion
ASH	Ashes (%)
BMP	Bio-Methane Potential
BOD	Biochemical oxygen demand (mg/L)
C/N	Carbon to Nitrogen ratio
F.C.	Fixed Carbon (%)

FVW	Fruit and Vegetable Wastes
KW	Kitchen Waste
LCFA	Long chain fatty acids
M	Moisture (%)
n.r.	not reported
OFMSW	Organic Fraction Of Municipal Solid Waste
PS	Primary Sludge
SPG	Specific gas production (Nm_3/kg VS)
SRT	Solids retention time (days)
TAS	Thickened Activated Sludge
TS	The total content of solids (%)
VS	Volatile solids (%)
WAS	Waste of Activated Sludge
WUS	Waste Untreated Sludge

References

1. Council Directive. 1999/31/EC of 26 April 1999 on the landfill of waste. *Off. J. Eur. Commun.* **1999**, *L182*, 1–19.
2. Murto, L.; Bjornsson, B.; Mattiasson, B. Impact of food industrial waste on anaerobic co-digestion of sewage sludge and pig manure. *J. Environ. Manag.* **2004**, *70*, 101–107.
3. European Commission. *Communication from the Commission to the Council and the European Parliament on Future Steps in Bio-Waste Management in the European Union*; COM235 Final; European Commission: Brussels, Belgium, 2010.
4. European Commission. *Environmental, Economic and Social Impacts of the Use of Sewage Sludge on Land*; Final Report, Part III: Project Interim Reports; European Commission: Brussels, Belgium, 2010.
5. European Commission. *European Commission Environmental Statistics and Accounts in Europe*; European Commission: Brussels, Belgium, 2010.
6. Czepiel, P.M.; Crill, P.M.; Harriss, R.C. Methane Emissions from Municipal Wastewater Treatment Processes. *Environ. Sci. Technol.* **1993**, *27*, 2472–2477.
7. Kolbl, S.; Forte-Tavčer, P.; Stres, B. Potential for valorization of dehydrated paper pulp sludge for biogas production: Addition of selected hydrolytic enzymes in semi-continuous anaerobic digestion assays. *Energy* **2017**, *126*, 326–334.
8. Kolbl, S.; Paloczi, A.; Panjan, J.; Stres, B. Addressing case specific biogas plant tasks: Industry oriented methane yields derived from 5 L Automatic Methane Potential Test Systems in batch or semi-continuous tests using realistic inocula, substrate particle sizes and organic loading. *Bioresour. Technol.* **2014**, *153*, 180–188.
9. Cabbai, V.; Ballico, M.; Aneggi, E.; Goi, D. BMP tests of source selected OFMSW to evaluate anaerobic codigestion with sewage sludge. *Waste Manag.* **2013**, *33*, 1626–1632.
10. Bolzonella, D.; Battistoni, P.; Susini, C.; Cecchi, F. Anaerobic codigestion of waste activated sludge and OFMSW: The experiences of Viareggio and Treviso plants (Italy). *Water Sci. Technol.* **2006**, *53*, 203–211.
11. Cavinato, C.; Bolzonella, D.; Pava, P.; Fatone, F.; Cecchi, F. Mesophilic and thermophilic anaerobic co-digestion of waste active sludge and source sorted biowaste in pilot and full scale reactors. *Renew. Energy* **2013**, *55*, 260–265.
12. Nielfa, A.; Cano, R.; Fdz–Polanco, M. Theoretical methane production generated by the co-digestion of organic fraction municipal solidwaste and biological sludge. *Biotechnol. Rep.* **2015**, *5*, 14–21.
13. Kim, H.W.; Han, S.K.; Shin, H.S. The optimization of food waste addition as aco-substrate in anaerobic digestion of sewage sludge. *Waste Manag. Res.* **2003**, *21*, 515–526.
14. Gomez Lahoz, C.; Fernandez Gimenez, B.; Garcia Herruzo, F.; Rodriguez Maroto, J.M.; Vereda-Alonso, C. Biomethanization of mixtures of fruits and vegetables solid wastes and sludge from a municipal wastewater treatment plant. *J. Environ. Sci. Health A Tox. Hazard Subst. Environ. Eng.* **2007**, *42*, 481–487.
15. Sosnowski, P.; Wieczorek, A.; Ledakowicz, S. Anaerobic codigestion of sewage sludge and organic fraction of municipal solid waste. *Adv. Environ. Res.* **2003**, *7*, 609–613.
16. Zitomer, D.H.; Johnson, C.C.; Speece, R.E. Metal Stimulation and Municipal Digester Thermophilic/Mesophilic Activity. *J. Environ. Eng.* **2008**, *134*, 42–47.

17. El Zein, A.; Seif, H.; Gooda, E. Effect of Co-composting Fish and Banana Wastes with Organic Municipal Solid Wastes on Carbon/Nitrogen Ratio. *Civ. Environ.* **2015**, *7*, 122–139.

18. Kaparaju, P.; Ellegaard, L.; Angelidaki, I. Effects of mixing on methane production during thermophilic anaerobic digestion of manure: Lab-scale and pilot-scale studies. *Bioresour. Technol.* **2008**, *99*, 4919–4928.

19. Elsayed, M.; Andres. Y.; Blel, M.; Gad, A. Methane Production By Anaerobic Co-Digestion of Sewage Sludge and Wheat Straw Under Mesophilic Conditions. *Int. J. Sci. Technol. Res.* **2015**, *4*, 1–6.

20. Sreekrishnan, T.R.; Kohli, S.; Rana, V. Enhancement of biogas production from solid substrates using different techniques—A review. *Bioresour. Technol.* **2004**, *95*, 1–10.

21. Mata-Alvarez, J.; Dosta, J.; Macè, S.; Astals, S. Codigestion of solid wastes: A review of its uses and perspectives including modelling. *Crit. Rev. Biotechnol.* **2011**, *31*, 99–111.

22. Pereira, M.A.; Pires, O.C.; Mota, M.; Alves, M.M. Anaerobic biodegradation of oleic and palmitic acids: Evidence of mass transfer limitation caused by long chain fatty acid accumulation onto the anaerobic sludge. *Biotechnol. Bioeng.* **2005**, *92*, 15–23.

23. Pereira, M.A.; Cavaleiro, A.J.; Mota, M.; Alves, M.M. Accumulation of long-chain fatty acids onto anaerobic sludge under steady state and shock loading conditions: Effect on acetogenic and methanogenic activity. *Water Sci. Technol.* **2003**, *48*, 33–40.

24. Zaher, U.; Li, R.; Jeppsson, U.; Steyer, J.P.; Chen, S. GISCOD: General integrated solid waste co-digestion model. *Water Res.* **2009**, *43*, 2717–2727.

25. Bolzonella, D.; Battistoni, P.; Mata-Alvarez, J.; Cecchi, F. Anaerobic digestion of organic solid wastes: Process behaviour in transient conditions. *Water Sci. Technol.* **2003**, *48*, 1–8.

26. Caffaz, S.; Bettazzi, E.; Scaglione, D.; Lubello, C. An integrated approach in a municipal WWTP: Anaerobic codigestion of sludge with organic waste and nutrient removal from supernatant. *Water Sci. Technol.* **2008**, *58*, 669–676.

27. Zupančiča, D.G.; Uranjek-Ževartb, N.; Roša, M. Full-scale anaerobic co-digestion of organic waste and municipal sludge. *Biomass Bioenergy* **2008**, *32*, 162–167.

28. Buratti, C.; Barbanera, M.; Bartocci, P.; Fantozzi, F. Thermogravimetric analysis of the behavior of sub-bituminous coal and cellulosic ethanol residue during co-combustion. *Bioresour. Technol.* **2015**, *186*, 154–162.

29. Fantozzi, F.; Buratti, C.; Morlino, C.; Massoli, S. Analysis of biogas yield and quality produced by anaerobic digestion of different combination of biomass and inoculums. In Proceedings of the 16th Biomass Conference and Exhibition, Valencia, Italy, 2–4 June 2008.

30. Fantozzi, F.; Buratti, C. Biogas production from different substrates in an experimental continuously stirred tank reactor anaerobic digester. *Bioresour. Technol.* **2009**, *100*, 2783–5789.

31. Fantozzi, F.; Buratti, C. Anaerobic digestion of mechanically treated OFMSW: Experimental data on biogas/methane production and residues characterization. *Bioresour. Technol.* **2011**, *102*, 8885–8892.

32. Ismail, Z.Z.; Talib, A.R. Assessment of anaerobic co-digestion of agro wastes for biogas recovery: A bench scale application to date palm wastes. *Energy Environ.* **2014**, *5*, 591–600.

33. Jianzheng, L.; Ajay, K.; Junguo, H.; Qiaoying, B.; Sheng, C.; Peng, W. Assessment of the effects of dry anaerobic co-digestion of cow dung with waste water sludge on biogas yield and biodegradability. *Int. J. Phys. Sci.* **2011**, *5*, 591–600.

34. Yao, F.X.; Macías, F.; Santesteban, A.; Virgel, S.; Blanco, F.; Jiang, X.; Camps Arbestain, M. Influence of the acid buffering capacity of different types of Technosols on the chemistry of their leachates. *Chemosphere* **2009**, *74*, 250–258.

35. Fonoll, X.; Astals, S.; Dosta, J.; Mata-Alvarez, J. Anaerobic co-digestion of sewage sludge and fruit wastes: Evaluation of the transitory states when the co-substrateis change. *Chemosphere* **2015**, *262*, 1268–1274.

36. Zhang, P.; Zhang, G.; Wang, W. Ultrasonic treatment of biological sludge: Floc disintegration, cell lysis and inactivation. *Bioresour. Technol.* **2007**, *98*, 207–210.

37. Lebiocka, M.; Piotrowicz, A. Co-digestion of sewage sludge and organic fraction of municipal solid waste. Acomperison between laboratory and technical scales. *Environ. Prot. Eng.* **2012**, *38*, 157–162.

38. Borowski, S. Co/digestion of the hydromechanically separated organic fraction of municipal solid waste with sewage sludge. *J. Environ. Manag.* **2015**, *147*, 87–94.

39. Heo, N.H.; Park, S.C.; Kang, H. Effects of mixture ratio and hydraulic retention time on single-stage anaerobic co-digestion of food waste and waste activated sludge. *J. Environ. Sci. Health A Tox. Hazard Subst. Environ. Eng.* **2004**, *39*, 1739–1756.

40. Gomez, X.; Cuetos, M.J.; Cara, J.; Moran, A.; Garcia, A.I. Anaerobic co-digestion of primary sludge and the fruit and vegetable fraction of the municipal solid wastes: Conditions for mixing and evaluation of the organic loading rate. *Renew. Energy* **2006**, *31*, 2017–2024.

fermentation

MDPI

Article

Optimization of the Enzymatic Saccharification Process of Milled Orange Wastes

Daniel Velasco, Juan J. Senit, Isabel de la Torre, Tamara M. Santos, Pedro Yustos, Victoria E. Santos and Miguel Ladero *

Chemical Engineering Department, Chemistry College, Complutense University, 28040 Madrid, Spain; danielvc_iq@hotmail.com (D.V.); juanjo.senit@gmail.com (J.J.S.); itpascual@ucm.es (I.d.l.T.); tmsantos@quim.ucm.es (T.M.S.); pyustosc@ucm.es (P.Y.); vesantos@ucm.es (V.E.S.)
* Correspondence: mladerog@ucm.es; Tel.: +34-91-394-4164

Received: 2 July 2017; Accepted: 30 July 2017; Published: 1 August 2017

Abstract: Orange juice production generates a very high quantity of residues (Orange Peel Waste or OPW-50–60% of total weight) that can be used for cattle feed as well as feedstock for the extraction or production of essential oils, pectin and nutraceutics and several monosaccharides by saccharification, inversion and enzyme-aided extraction. As in all solid wastes, simple pretreatments can enhance these processes. In this study, hydrothermal pretreatments and knife milling have been analyzed with enzyme saccharification at different dry solid contents as the selection test: simple knife milling seemed more appropriate, as no added pretreatment resulted in better final glucose yields. A Taguchi optimization study on dry solid to liquid content and the composition of the enzymatic cocktail was undertaken. The amounts of enzymatic preparations were set to reduce their impact on the economy of the process; however, as expected, the highest amounts resulted in the best yields to glucose and other monomers. Interestingly, the highest content in solid to liquid (11.5% on dry basis) rendered the best yields. Additionally, in search for process economy with high yields, operational conditions were set: medium amounts of hemicellulases, polygalacturonases and β-glucosidases. Finally, a fractal kinetic modelling of results for all products from the saccharification process indicated very high activities resulting in the liberation of glucose, fructose and xylose, and very low activities to arabinose and galactose. High activity on pectin was also observed, but, for all monomers liberated initially at a fast rate, high hindrances appeared during the saccharification process.

Keywords: biorefinery; saccharification; orange waste; valorization; optimization; Taguchi design

1. Introduction

The integrated production of chemicals, fuels, food and feed, as well as thermal energy and electricity, using biomass as a renewable resource is the aim of a biorefinery [1]. Biomass is a plentiful resource, as it includes wood and its residues, agricultural crops and their corresponding residues, food waste, municipal solid waste as well as algae and microalgae. Biomass production is able to fix up to 0.02% of the incident sun energy, while reducing CO_2 content in the atmosphere through photosynthesis, thus ensuring, from a theoretical point of view, a progressive swift from fossil resources to the original material behind their formation: biomass [2,3]. At the same time, second generation biorefineries are based mainly in renewable non-food resources, such as lignocellulosic biomass [4]. This abundant biomass conversion into chemicals and fuel poses several challenges that currently jeopardize its implementation due to economic reasons [5].

The increasing demand for food and feed due to the demographic surge in the last decades of the 20th century and during this century has created an enormous amount of food residues. According to the Food and Agriculture Organization of the United Nations (FAO), about 30% of food produced is lost during harvesting, manufacturing, and household and industrial consumption and disposal, amounting to over 1300 million tonnes [6]. Lal et al. [7] estimated the energy content for crop residues in *circa* 7.5 billion barrels (bbl) of diesel–estimation for 2005 that, if revised for production in 2014, would be increased by 23%, considering data from FAOSTAT [8]. The most abundant fruit tree in the world is sweet orange (*Citrus sinensis* L.), whose harvest creates almost 60% of the total world citrus fruits production (slightly over 70 million tonnes per year) [9]. Its production and transformation into juice results in almost 30 MM tonnes/year of peel, pulp and seeds waste (OPW). Part of it can be included in cattle feed [10] but, despite this use, such an amount is more than enough to search for further applications of this waste. Some applications in the literature include the use as bioadsorbent for site remediation and wastewater treatment, as raw material for biorefineries (for the production of essential oils, pectin, bioethanol, energy, food added-value ingredients including nutraceuticals, dietary fiber, carotenoids, vitamins, etc.), in the production of electricity by microbial fuel cells or isolation of essential oils [9,11–18].

Residues from orange juice factories are formed by orange pulp, seeds and peel. This peel is composed of the outer orange layer (flavedo) and the white and spongy inner layer (albedo). Orange pulp is very humid and rich in monosaccharides (glucose and fructose) and disaccharides (sucrose), as it is composed by vesicles containing the juice. The albedo layer is rich in pectin, while the flavedo layer contains a good amount of essential oils, with limonene as the main compound in them, and flavonoids. In short, orange waste contains low levels of lignin and protein, medium levels of cellulose and hemicellulose and high levels of pectin. Although their composition can change with the cultivar, the season of the year, the country and the technology used for juice extraction, all these residues are very humid, with a water content ranging from 80% to 84% [16,19].

In the biochemical approach of biorefineries, the two most important processes ensuring their economic success are the raw material pretreatment and fractionation, and the enzymatic saccharification [20]. The pretreated material should both be accessible to enzymes and very reactive, so a good pretreatment will reduce the content of inert or hindering biopolymers (lignin, hemicellulose, pectin) while increasing porosity, hydration, external surface, and the amorphous nature of cellulose [21]. Classical pretreatments have been tested with orange waste: acid and basic treatments, steam injection, steam explosion, etc., to observe that mild conditions of pressure, concentration and temperature suffice to render a material with high reactivity [22–25]. In fact, the combination of thermal, mechanical and chemical pretreatments result in a very high removal of pectin, thus aiding in the subsequent saccharification process [23]. Enzymatic saccharification is a combination of hydrolysis reactions resulting in the depolymerization of polisaccharides and the final production of monosaccharides, with glucose as the main target in most cases. This process is very complex due to its multiphasic nature, with several solid phases (if different polymers in the lignocellulosic matrix are regarded as such) interacting with a liquid phase, and several enzymes catalyzing reactions either in the solid surface or in the liquid phase. Several phenomena hinder these hydrolytic steps, creating an initial burst phase followed by a deep reduction in overall rates of monosaccharide production, due to a series of interfering phenomena that gain importance along the process. At the same time, high dry solid concentrations and low enzyme amounts are sought to reduce the cost impact of this time consuming process while reaching high titers for glucose and other monosaccharides [26–29]. To reduce the reduction of overall rates to product, several operational factors should be analyzed and optimized: mixing, agitation, adequate pretreatment(s), more active enzyme cocktails, avoidance or reduction of non-producing enzyme-solid interactions, and more [29–31]. Regarding orange waste, some papers report on the optimization of the pretreatment stage [32] or pectin extraction [33], but, to the knowledge of the authors, no paper is devoted to the optimization of the saccharification process.

This paper focuses on the optimization of the enzyme cocktail and the dry solid to liquid ratio for a partially dried and milled orange waste from juice factories. The optimization is performed by using a Taguchi design of experiments, after some previous experiments were developed to choose the more adequate pretreatment from a techno-economic point-of-view. Adequate conditions result in very high yields and good productivities to glucose and fructose at moderate enzyme loadings and medium solid to liquid ratios.

2. Materials and Methods

2.1. Materials

The orange waste (OPW) was provided by Biopolis S.L. (Valencia, Spain). A first pretreatment after their reception was knife-milling with a Knife Mill Grindomix GM 200 (Retsch, Haan, Germany), followed by sieving to a final average particle diameter of 2.2 ± 0.4 mm. This fraction was kept frozen at $-20\,^{\circ}C$ until the moment of use.

The enzymatic cocktail consisted of a mixture of several industrial products kindly provided by Novozymes: Celluclast 1.5 L, Novozym 188, and Pectinex Ultra SP-L, containing glucanases, polygalacturonases, β-glucosidases, xylanases, β-xylosidases and several auxiliary activities. Their combined activities drive the hydrolysis of cellulose, hemicelluloses and pectin in parallel.

Several substrates for testing enzymatic activities were provided by Sigma-Aldrich (Saint Louis, MO, USA): Filter paper Whatman 1, Avicel PH-101, polygalacturonic acid and ONPG (*o*-nitrophenyl-β-D-glucoside). Citric acid, NaOH, and HCl for buffer solutions and pH adjustment were from Sigma-Aldrich, as all other reagents needed, all of them of reagent or better grade. Monosaccharides (glucose, fructose, galactose, xylose and arabinose) and galacturonic acid were purchased from the same supplier and employed as per HPLC analytical standards.

2.2. Methods

2.2.1. Basic Chemical Analysis of Solids

The basic chemical characterization was performed following National Renewable Energy Laboratory (NREL) USA procedures for the determination of glucans, arabinans, galactans, xylans, polygalacturonic acid, citric acid and fructose, apart for a previous analysis of extractives, ash and moisture (infrared drying at $70–90\,^{\circ}C$ till constant weight) [34–36]. These procedures were applied not only to the original material, but also to the materials resulting from several thermal pretreatments applied to the milled fraction.

2.2.2. Determination of Enzyme Activities

The activities of the main enzymes were tested, in triplicates, with several model substrates: global cellulase activity was measured on Whatman filter paper 1; the cellobiohydrolase activity was measured on Avicel PH-101; *o*-nitrophenyl-β-D-glucopyranoside was the substrate for β-glucosidase testing, while polygalacturonic acid was used for pectin depolymerase (PD) activity. In all cases, a 50 mM citrate buffer pH 4.8 and a temperature of $50\,^{\circ}C$ were the fixed operational conditions and a well-stirred batch reactor was employed. Several samples were withdrawn throughout the progress of the reaction and analyzed by HPLC, as explained in a subsequent section. The testing procedures are described in several references [37,38]. The amount of protein in each saccharification run was analyzed by the Bradford method, with BSA (Bovine Serum Albumin) as standard and Coomassie Brilliant Blue G250 as the dye [39].

2.2.3. Hydrothermal Pretreatments

The milled and sieved orange waste was subjected to either partial drying at $70\,^{\circ}C$ down 60% humidity or, as a previous process, to hydrodistillation (for several hours), microwave water extraction and steam stripping (for several minutes) to remove essential oils. All these pretreatments were performed at 100 to $120\,^{\circ}C$ at near atmospheric pressure. For hydrodistillation, a Clevenger apparatus was fitted to a round-bottom flask and a Dimroth condenser. For microwave heating, a 500 W power was set in a Milestone EthosX (Milan, Italy). Final solids were extracted with hexane (1% w/w dry solid/liquid) to analyze their content in essential oils by GC-MS. GC-MS analysis was performed in a GC-2010 Shimadzu with a single quadrupole detector GCMS-QP2010 Plus containing a ZB 1-ms column (10 m × 0.1 mm × 0.1 mm) using 99.999% v/v helium as eluent with a ramp from 160 to 200 °C at 5 °C/min and a final period of 15 min at 200 °C.

2.2.4. Saccharification Experiments

Saccharification runs were carried out in triplicates, in thermostated 50 mL round-bottom flasks with magnetic agitation and placed in thermostated aluminum blocks. In each run, 6 g of solids with 60% humidity produced by the several pretreatments tested were employed. Depending on the pretreatment, either 20 mL of citrate buffer solution 50 mM pH 5.2 were used or citric acid was added as needed (after pretreatment liquid phase HPLC analysis) and pH readjusted to 5.2. The total percentage of dried solids changed from 6.7% to 12.5% depending on the conditions to be tested for pretreatment comparison or for saccharification optimization. In all cases, the pH was controlled at 5.2 by adding NaOH 5 M after the addition of the enzymes during the first hour (burst phase), until a low viscosity was obtained.

In each saccharification run and in the case of the liquid phase after some pretreatments, several samples were withdrawn with reaction time. Sample preparation included centrifugation, dilution and filtration. The resulting liquid samples were analyzed by HPLC with JASCO 2000 series equipment (Tokyo, Japan), while analyte detection was performed with a refraction index detector. Acidified Milli-Q water (0.005 N H_2SO_4) flowing at 0.5 mL/min was employed as mobile phase, while the stationary phase was a Rezex ROA-Organic Acid H^+ (8%) column (300 mm × 7.80 mm) placed in an oven at 60 °C. For the determination of xylose, arabinose and galactose, a Rezex RCM-Monosaccharide Ca^{2+} column (300 mm × 7.80 mm) at 80 °C was employed, with Milli-Q water as eluent (Both columns from Phenomenex, Torrance, CA, USA).

The comparison of the results obtained by HPLC analysis with original or pretreated solid hydrolyzed by using NREL procedures and enzymatic saccharification (in this case, at several reaction times) was employed as the basis to calculate yields. A total yield (Y_T) defines the amount of total glucose released by all possible means: enzyme aided extraction, sucrose inversion and saccharification, while the saccharification yield (Y_S) can be established by knowing the fructose released, as this monosaccharide proceeds only from inversion and enzyme-aided extraction, and its ratio to free glucose is well-known (average fru/glu in orange juice = 1.025).

$$Y_T(\%) = \frac{C_{\text{total glu saccharification}}}{C_{\text{total glu released by NREL procedure}}} \times 100 \tag{1}$$

$$Y_S(\%) = \frac{C_{\text{total glu saccharification}} - C_{\text{total fru saccharification}}/1.025}{C_{\text{cellulose}} \times 1.1} \times 100 \tag{2}$$

2.2.5. Statistical Design of Experiments

A fractionated factorial design known as the Taguchi method was employed as a Design of Experiments (DOE) technique, which is based on: an orthogonal fractionated factorial design methodology. The application of this methodology results in a dramatic reduction of the number of experiments to be performed [40–42].

For enzymatic saccharification, there are many variables to be considered, although a pH of 5.0 or temperature of 50 °C are well established in the literature. As for other factors, like mixing or agitation, these should be fixed at medium to high values to avoid hindrances. To reach high concentrations of monosaccharides at an acceptable amount of enzymes, the factors to consider are the dry solid to liquid ratio, and the amount of the three industrial preparations used here: Celluclast 1.5 L, Novozym 188 and Pectinex Ultra-SP L.

Taguchi methodology has been followed according to the next steps: (1) selection of the responses or dependent variables to be optimized (in this case, total and saccharification yields to glucose);

(2) identification of the factors and the choice of their values or levels; (3) performance of duplicate saccharification runs; (4) statistical analysis of results and the signal-to-noise ratio, determining the best values or levels for each factor and predicting the result for the optimal run; (5) triplicate conduction (of the confirmatory run) in the optimal conditions.

As concentrations and yields have to be maximized, the signal-to-noise ratio (S/N ratio or SNR) to be calculated in this case is the higher-the-better one, by using the following expression:

$$S/N \text{ ratio} = -10 \, \log_{10} \frac{1}{n} \sum_{i=1}^{n} \frac{1}{y_i^2} \tag{3}$$

where y_i is the value of the response or objective variable (yield, in this case) for each run and n is the number of repetitions per run. To examine the importance of every single individual factor on the saccharification, the average S/N for each factor at each level is computed with equation [4].

$$S/N = \frac{\text{Sum of } S/N \text{ values for factor } i \text{ at each level } j}{3} \tag{4}$$

The interval of S/N ratios is calculated for each factor and compared to a global average value for all factors. The wider the interval the more influential the factor on the objective variable is. Moreover, given the optimal levels for all factors, the mean response predicted for the best conditions can be calculated with the next equation (for factors A, B, C, D):

$$\text{Best yield predicted} = Y + (A_{\text{opt}} - Y) + (B_{\text{opt}} - Y) + (C_{\text{opt}} - Y) + (D_{\text{opt}} - Y) \tag{5}$$

where Y is the global average value for each yield (total, saccharification).

2.2.6. Kinetic Modelling

Fractal kinetic models are semi-empirical models that take into account the severe reduction in global productivity or rate in saccharification processes. The solids on which several of the enzymes involved are acting are similar to fractals in the sense that not all the surface is available for enzymatic action; therefore, there is no integer dimension at any reaction time and the solid is a fractal. This fractal nature extends to dynamic processes, as the controlling nature of each kinetic or dynamic phenomenon interacting during the whole process can change during saccharification [43–45]. Fractal models assume that hindrance increases with time, so kinetic constants decrease in the same direction, meaning that they are a function of time. The simplest fractal kinetic model supposes that the first hydrolysis steps are of first-order, as water is in excess.

$$R_G' = \frac{dX}{dt} = k' \cdot (1 - X) \tag{6}$$

where the kinetic constant k' for the first hydrolytic event is a first-order constant and, for the subsequent hydrolysis, a function of time, according to:

$$k' = k \cdot t^{-h} \, \forall \, t \geq 1 \text{ or } k' = k \, \forall \, 0 \leq t \leq 1 \tag{7}$$

Therefore, for the very first hydrolysis taking place on the surface, the following equation is valid. After fitting the model to every set of kinetic data, goodness-of-fit can be expressed by several parameters: the sum of quadratic residues (SQR) (which should be near or equal to zero), Root-Mean-Square-Error ($RMSE$), the square root (SQR) divided by the degrees of freedom (same trend or value as SSR), and Fisher F (it should be high, as it includes SQR in the denominator). They can be calculated with these equations:

$$SQR = \sum_{i=1}^{N} \left(y_{i,\text{exp}} - y_{i,\text{calc}} \right)^2 \tag{8}$$

$$RMSE = \sqrt{\frac{SQR}{N-K}} \tag{9}$$

$$F = \frac{\sum_{i=1}^{N} \left(y_{i,\text{calc}} \right)^2 / K}{\sum_{i=1}^{N} SQR / (N-K)} \tag{10}$$

In these equations, $y_{i,\text{exp}}$ is the conversion of each product after the maximum concentration of that monosaccharide is estimated by NREL procedures, $y_{i,\text{calc}}$ is the same conversion, but calculated with the model and the optimal values of the kinetic parameters, N is the number of data and K, the number of kinetic/thermodynamic constants in the model (2, for the model proposed).

3. Results and Discussion

3.1. Preliminary Experiments: Hydrothermal Pretreatments

Prior to the performance of saccharification runs, the protocols established in NREL were used to determine the basic composition of several stocks of OPW pretreated by mild water thermal procedures or, in their absence, by mechanical knife milling (at least). Table 1 compiles the results and, as can be seen, water content depended on the pretreatment time as well as the physical state of the water and the type of device used to heat the solid (indirect electrical heating or direct microwave heating). Direct heating or steaming pretreatments result in higher hydration of the solid in only a few minutes. However, indirect or classical heating involves a very long time to increase water content in OPW. The more aggressive the hydration procedure the more evident the presence of hydrolysis reactions is within the solids, with an increase of the soluble part of any of the polymers considered. The original OPW contains a high amount of water, no extractives in chloroform and a high percentage of extractives in solids (38.5% ± 1.2%); pretreatments with steam or hot water lead to a reduction in free sugars and an increase of polymers (in mass percentage) compared to the original solid. Again, this extraction of free sugars is more effective when using microwaves or direct steaming (followed by rinsing with water in a Buchner funnel).

Table 1. Compositional analysis of original and pretreated solids based on NREL methodology (National Renewable Energy Laboratory, US Department of Energy).

OPW Pretreatment	Solid Dry Weight	Glucan Sol/Ins	Xylan Sol/Ins	Other Sugars Sol/Ins	Pectin Sol/Ins	Lignin/Ash
Knife Milling (KM)	16.5 ± 0.2	5.12/12.9	0.04/0.72	1.91/13.5	0/18.6	6.47/3.72
KM+ 6 h LHW 1.2 atm 120 °C	15.7 ± 0.4	6.41/13.7	0.06/0.81	2.56/13.2	2.42/17.2	7.21/3.96
KM+ 24 h LHW 1.2 atm 120 °C	15.1 ± 0.2	6.96/14.4	0.06/0.84	4.56/12.9	4.56/14.2	7.42/4.01
KM+ 40 min MW 500 w 1.2 atm 120 °C	12.1 ± 0.3	7.89/15.4	0.45/1.12	7.56/8.45	7.84/12.2	8.12/4.29
KM+ 80 min MW 500 w 1.2 atm 120 °C	10.9 ± 0.1	8.95/17.1	0.89/1.04	10.2/4.68	9.41/8.56	8.23/4.56
KM+ 40 min direct steaming	10.2 ± 0.4	10.6/11.5	1.12/0.98	10.5/4.21	9.24/8.69	9.12/5.14

All results in % or % dry solid; Sol/Ins = soluble/insoluble; atm = atmosphere (pressure).

After plain knife milling or followed by hot water or steaming pretreatments, solids were partially dried to a 70% water content, so as to avoid a very high volume of solids for the next step and, at the same time, the collapse of their porous structure. Saccharification runs were performed at a relatively high content of cellulases (Celluclast 1.5 L), β-glucosidase and xylanases/β-xylosidases (Novozym 188) and polygalacturonases (PG) and auxiliary enzymes (Pectinex Ultra SP-L): 5.8 mg total protein per gram dry solid (DS), 98 FPU/g DS, 199 UI β-glu/g DS, and 11400 UI PG UI/g DS. With these conditions and high stirring at 500 rpm, a 5.0 pH value and 50 °C applied during the 72 h of the hydrolytic process, several effects due to mass transfer and low activity can be avoided, and results depend mainly on the pretreatment employed in each case. These are shown in Figure 1, as total yield to glucose, on one hand, at total liberated glucose concentration, in g/L, on the other.

A very mild mechanical pretreatment, knife milling, renders a solid very reactive, while all other pretreatments, except steaming, induce a considerable reduction in global yields and final glucose concentrations. Since the solids, by their soluble polymer content, should be more reactive, this situation should be due to a collapse of the porous structure, resulting in a mass transfer hindrance of the enzymes into the structure. In fact, if steaming is employed, it is reasonable to accept that this structure is not being affected by the treatment in this sense, but probably otherwise, leading to swelling. However, the energy used for this further pretreatment hardly compensates for only knife milling, at least, until it reaches medium to high dry solid content (15%). Therefore, knife milling followed by partial drying to 70% water content is the pretreatment used for further experimentation, reducing energy consumption to low values due to the softness of OPW.

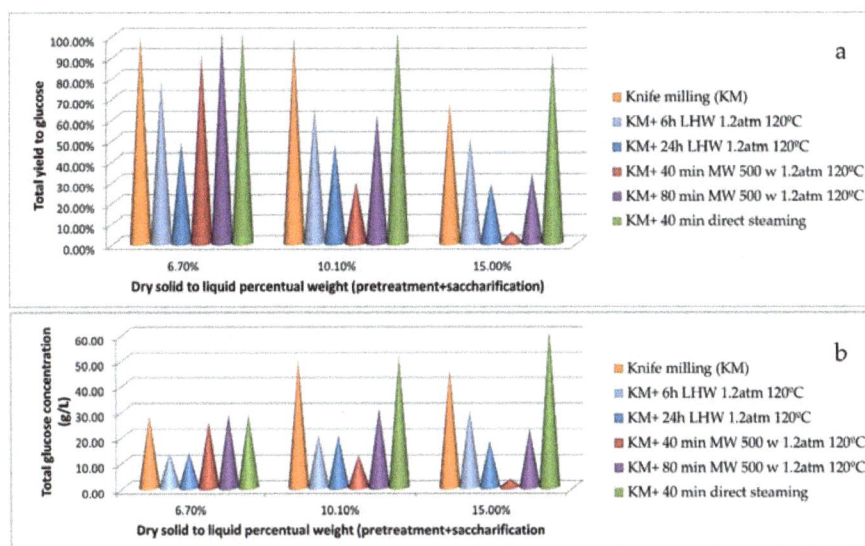

Figure 1. Results for deep enzymatic saccharification for various OPW (Orange Waste) solids obtained by several mechanical and hydrothermal pretreatments: (**a**) Percentual total yield (Y_T) to glucose at 72 h; (**b**) Final glucose concentration liberated at 72 h.

3.2. Taguchi Optimization of the Saccharification

The Taguchi method applied in this case focused on total and saccharification yields optimization, with a higher emphasis on the former, as a high concentration of monosaccharide is important, regardless of its origin. The number of factors or independent variables is four and three levels or values are chosen for each of them, as shown in Table 2. With these conditions, nine runs are needed to cover the experimental range under study (18 if duplicate experiments are performed). In general, the amount of runs N is a function of the number of factors P and the number of levels chosen for them L, as shown in Equation (11). This strategy avoids any bias in the selection of runs and in the information extracted from them.

$$N = (L-1)P + 1 \tag{11}$$

As for the total maximum concentration of protein (1.16 mg per gram of dry solid), it is evident that low concentrations of enzymes are selected to avoid a high cost in saccharification. At the same time, partial drying permits the increment of dry solid per liquid, so as to approach the solid amount usually found in wood and herbaceous raw materials saccharifications (from 15% upwards) and reach high concentrations of monosaccharides, if the final yields are favorable.

Table 3 displays the results of the saccharification runs after 72 h of process. At first glance, the increment of solid is not deleterious to the process owing to the narrow range of dry solid to liquid weight ratios here studied. The overall concentrations of glucose and fructose are medium to high, hence indicating good progress of enzyme-aided extractions, inversion and saccharification processes; however, the amount of sugars coming from hemicelluloses (arabinose, galactose and xylose) is relatively low, showing a certain recalcitrance of this polymer to the action of hemicellulases in Novozym 188. This can be expected as this preparation is more effective in wood and herbaceous hemicellulases, rich in xylose and mannose (and, in some cases, in glucose).

Table 2. L9 Taguchi matrix for exploring the experimental range of dry solid to liquid ratio and the diverse enzymatic activities included in the final enzymatic cocktail.

Run	% DS		Cellulase FPU/g DS		β-Glucosidase (UI)/g DS		Polygalacturonase UI/g Pectin		Protein/Dry Solid mg/g
1-1'	6.1	*1*	6.08	*1*	12	*1*	670	*1*	0.29
2-2'	6.1	*1*	12.15	*2*	23	*2*	1340	*2*	0.58
3-3'	6.1	*1*	24.30	*3*	50	*3*	2850	*3*	1.16
4-4'	9.5	*2*	6.08	*1*	23	*2*	2850	*3*	0.40
5-5'	9.5	*2*	12.15	*2*	50	*3*	670	*1*	0.75
6-6'	9.5	*2*	24.30	*3*	12	*1*	1340	*2*	0.88
7-7'	11.5	*3*	6.08	*1*	50	*3*	670	*1*	0.57
8-8'	11.5	*3*	12.15	*2*	12	*1*	2850	*3*	0.51
9-9'	11.5	*3*	24.30	*3*	23	*2*	1340	*2*	0.96

Factor A: DS/L ratio; Factor B: Cellulase; Factor C: β-glucosidase; Factor D: Polygalacturonase. Each factor has three levels or values (in italics).

Table 3. L9 Taguchi matrix results in terms of monomer concentrations in the fluid phase and total and saccharification yield to glucose after 72 h.

Run	DP ≥ 2 (g/L)	Glucose (g/L)	Fructose (g/L)	Ara + gal + xyl (g/L)	Galacturonic Acid (g/L)	Y_T	Y_S
1-1'	1.30	14.56	11.56	1.82	4.52	50.90	10.72
2-2'	1.24	16.88	13.40	2.19	5.98	61.87	30.65
3-3'	1.01	22.64	18.30	4.27	9.28	82.95	69.01
4-4'	2.02	28.48	23.45	5.42	11.97	77.73	58.78
5-5'	0.90	33.11	24.74	3.68	9.40	89.89	81.62
6-6'	1.46	34.06	26.41	4.03	12.58	92.48	86.33
7-7'	1.35	45.02	32.12	5.32	12.94	99.68	99.42
8-8'	1.12	45.27	36.21	5.64	17.70	99.70	99.46
9-9'	2.70	41.87	32.73	4.89	7.89	92.75	86.82

After estimating the SNR for each case for the-higher-the-better case, taking into account that the higher the yield the better process performance is obtained, partial average *SNR* values for each factor and level (for example, average SNR values for the A factor, that is, dry solid to liquid ratio, is the average value calculated from the values estimated from runs 1, 2 and 3 for level 1 of this factor) and the overall average *SNR* values (for all 9 runs and duplicates) are represented. This is performed for all factors. Results are depicted in Figure 2 for the global yield to glucose and in Figure 3 for the saccharification yield, paying particular attention to showing the same interval for the ordinate axis, so as to compare the effects of each operational variable or factor on the yields.

Figure 2 indicates that the most influential factor is the ratio of dry solid to liquid, so the higher amount leads to higher yields, within the restricted range analyzed here. As expected, if higher amounts of enzyme are added, the global yield increases regardless of their activities, but this increment is more evident for glucanases than for pectinases, for example. In the case of β-glucosidase and hemicellulases, this increment is not so progressive than in the other types of enzymes, indicating a critical amount that enhances the process. In any case, the optimal situation includes the highest level for all cases, so the optimal run should be $A_3B_3C_3D_3$ if we consider the global yield as our target.

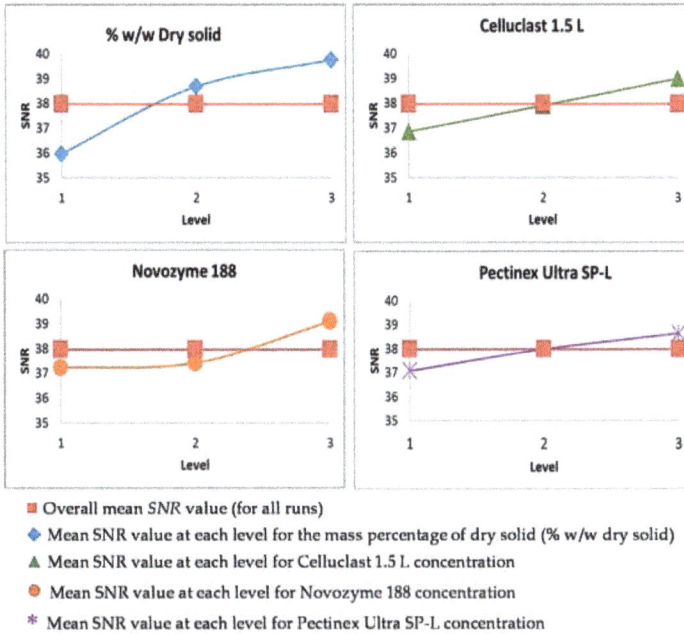

Figure 2. Taguchi plots (*SNR* vs. level for each factor) for the global yield to glucose.

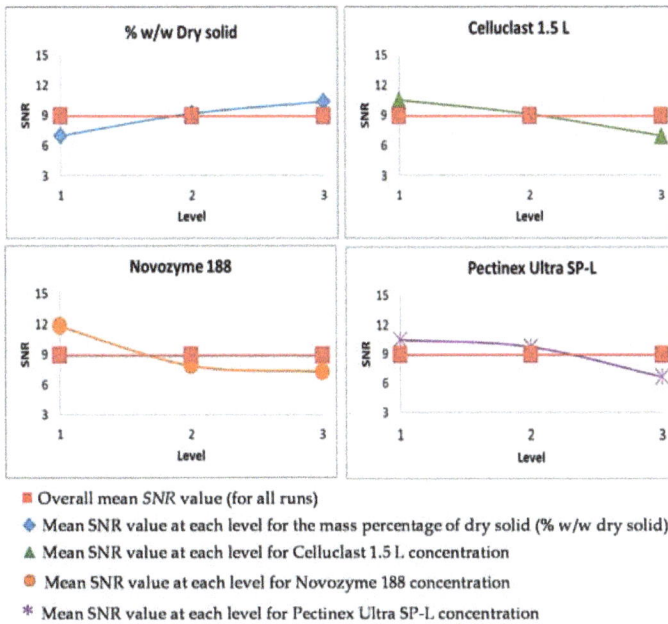

Figure 3. Taguchi plots (*SNR* vs. level for each factor) for the saccharification yield to glucose.

Saccharification yield was estimated by subtracting the effects of enzyme-aided extraction and sucrose inversion on the glucose concentration. The results obtained are graphed in Figure 3, showing that the effects are quite different from the previous case, although for the dry solid to liquid mass ratio, performance is similar. In all other cases, the lowest activity is better than the highest one. This observation can only be explained if an excess of protein results in non-active adsorption (non-active for saccharification purposes), maybe due to overcrowding, an effect that has previously been described for enzymes acting on lignocellulosic surfaces. For the saccharification yield, the best situation will be obtained with a $A_3B_1C_1D_1$ run.

Further analysis can be carried out by way of a technoeconomical assessment. For this purpose, a good benchmark can be a 42 HFCS (High Fructose Corn Syrup) 75–85% (content in solids) syrup, whose market price has steadily increased from 0.41 to 0.65 €/kg from January 2014 to May 2017, according to the Economic Research Service of the US Deparment of Agriculture (Sugar and Sweeteners Yearbook Tables). One should keep in mind that glucose and fructose are the main monosaccharides obtained in these saccharification runs with OPW as biomass substrate. Table 4 presents some estimations on glucose to fructose ratio and enzyme cost per kg mixture (prices provided by Novozymes for all preparations tested: Celluclast 1.5 L 20 €/kg; Novozym 188 35 €/kg; Pectinex Ultra SP-L 26 €/kg supplied in 25 kg drums–data provided by Novozymes-).

Table 4. L9 Taguchi matrix enzyme cost analysis with the glucose/fructose mixture as final monosaccharide mixture (compared to a 42 HFCS 75–85% syrup acting as benchmark).

Run	Enzyme Cost per Run	Glucose (g/L)	Fructose (g/L)	glu+fru (kg/10 L)	$\dfrac{\text{fru}}{\text{glu + fru}}$	Enzyme Cost (€/kg glu+fru)
1-1'	0.0468	14.56	11.56	0.2612	0.44	0.18
2-2'	0.1285	16.88	13.4	0.3028	0.44	0.42
3-3'	0.2570	22.64	18.3	0.4094	0.45	0.63
4-4'	0.2177	28.48	23.45	0.5193	0.45	0.42
5-5'	0.1894	33.11	24.74	0.5785	0.43	0.33
6-6'	0.1732	34.06	26.41	0.6047	0.44	0.29
7-7'	0.2445	45.02	32.12	0.7714	0.42	0.32
8-8'	0.2975	45.27	36.21	0.8148	0.44	0.37
9-9'	0.1691	41.87	32.73	0.746	0.44	0.23

It is evident that runs 7 and 8 are the best in this matrix if total sugars, total glucose or fructose concentrations are the objective functions to maximize. Although several cost effects are not included in this analysis, the conditions shown in run 9 appear much better if operational cost for the enzyme mixture amount, a well-known critical cost factor, is taken into account. Only run 1 is better than run 9 from this point of view, yet sugar amounts and yields are much worse in that case.

A final analysis and run performance applies Equation (5) for the chosen objective function or response. In this case, the global yield to glucose, being parallel to maximal concentrations of free monosaccharides, is the chosen one. In this case, the best conditions are $A_3B_3C_3D_3$ and the application of Equation (5) renders an expected global yield for this experiment of 108.2%. The validation runs give results of 99.85% and 99.64% global yield to glucose, which lead to an average final glucose concentration of 45.67 g/L, and 36.12 g/L of fructose, with an estimated enzyme cost of 0.51 €/kg of mixture glu+fru. Evidently, a very high yield to glucose (and other monosaccharides) results in a high cost, so the combination $A_3B_3C_2D_1$ (run 9) is much better from this perspective, with a good 92.75% global yield to glucose and a reduced cost (0.23 €/kg of mixture glu+fru). Even so, either further reductions in cost per kg industrial enzyme mixtures or higher specific activities at contained prices are needed to reduce the amount of protein (enzyme mixture) per gram dry solid.

3.3. Fractal Kinetic Modelling for Best Global Yield Conditions

A simple first-order fractal model has been fitted to kinetic data of all important monomers obtained in both corroboration runs (thus, optimal conditions regarding glucose total yield). Results on the kinetic parameters, with their standard errors, and goodness-of-fit parameters are provided in Table 5, while the fitting can be observed in Figure 4 in terms of total yields.

As may be seen from the kinetic parameters of the fractal equation for each case, it is perceptible that the standard error is low for most cases, thus indicating a good confidence degree in the optimal value of k' (the first-order kinetic constant) and h (the fractal exponent). The low values for SQR and RMSE and the high to very high values for Fisher F (much higher than the values tabulated at 95% confidence needed in all cases) indicate a good to very good fitting, an idea that is corroborated by Figure 4.

The values of the first-order kinetic constant show that the enzyme cocktail is mostly active on cellulose and a part of the hemicellulose, liberating glucose and xylose (not very abundant) at a fast pace. Fructose is free, as a monosaccharide, so it is liberated due to the disruption of the juice vesicles (as is the case of a high percentage of glucose). It is evident that this process is the fastest-occurring step. Regarding pectin hydrolysis, a high concentration of galacturonic acid is reached (more than 12 g/L) at 52 h, but its liberation is relatively slow, so PG activity seems to not be very high in this context. The same can be said for arabinose and galactose, the main monosaccharides in OPW hemicelluloses: they are liberated at a very slow rate and only partially in 52 h, so there are low activities for α-galactosidase and arabinan endo-1,5-α-L-arabinanase.

Looking at the fractal exponent, a high value indicates the presence on limiting phenomena hindering the liberation of the considered monomer. Thus, fructose liberation is fast at the beginning, but it is rapidly reduced, as shown by a high value of h for this monosaccharide. This fact suggests that the vesicles are being disrupted at a fast pace in the first moments, while fructose can be adsorbed in the solid part of the wastes and only the progressive disappearance of this part permits the liberation of a fraction of fructose. The same can be said for the free part of glucose, although in the case of xylose, being a minor component in the hemicellulosic fraction, the low liberation of galactose and arabinose can be a hindrance for xylose production. Finally, in the case of galacturonic acid, the exposure of pectin to the enzymatic action should be heterogeneous, as can be deduced from the complexity of pectin extraction in acid conditions, a very slow process that always leaves a good quantity of pectin in the solid matrix.

Table 5. Kinetic and goodness-of-fit parameters for the fractal models.

Parameter	Glucose	Fructose	Xylose	Galacturonic Acid	Arabinose	Galactose
k'	0.32 ± 0.03	0.47 ± 0.05	0.327 ± 0.021	0.116 ± 0.003	0.0098 ± 0.001	0.0092 ± 0.0003
h	0.62 ± 0.07	0.91 ± 0.09	1.11 ± 0.05	0.72 ± 0.01	0.15 ± 0.04	0.14 ± 0.01
SQR	0.0027	0.00112	0.00112	0.000096	0.00018	0.000016
RMSE	0.0198	0.021	0.0011	0.0037	0.005	0.0015
F-value	4294	3835	6420	41,795	3195	32,769

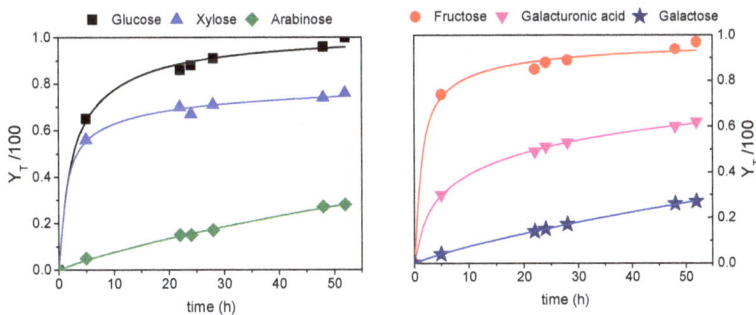

Figure 4. Fitting of the fractal models to the conversion data of monomers liberated during the saccharification process in optimal conditions to maximize the total glucose yield.

Fermentation **2017**, *3*, 37

4. Conclusions

This study indicates that knife milling, a mild mechanical pretreatment of orange peel waste, is enough to obtain acceptable yields for saccharification at low enzyme amounts. Furthermore, optimization of the hydrolysis and extraction enzymatic process performed on milled OPW shows that good to high yields of glucose and fructose are obtained, liberating more than 45 g/L glucose, more than 30 g/L fructose, and 12 g/L galacturonic acid within 52 h with a protein content under 1 mg per gram dry solid and at 11.5% w/w DS/L. Further enhancement of the enzymatic cocktail should be sought, especially to liberate the arabinose and galactose contained in the hemicellulose part of OPW. Reduction of the prices of pectinase and β-glu/hemicellulases industrial preparations should be pursued if the maximum yields to monomers are the main aim.

Acknowledgments: The kind gift of Celluclast 1.5 L, Novozym 188, and Pectinex Ultra SP-L by Ramiro Martinez (Novozymes Spain, Pozuelo de Alarcon—Madrid-Spain) is gratefully acknowledged. Likewise, the authors thank the supply of OPW sent by Marta Tortajada and Antonia Rojas (Biopolis S.L., Paterna—Valencia-Spain). This work was funded by MICINN under contracts CTQ-2013-45970-C2-1-R and PCIN-2013-021-C02-01.

Author Contributions: Miguel Ladero and Victoria E. Santos conceived and designed the experiments; Daniel Velasco, Juan J. Senit, Isabel de la Torre and Tamara M. Santos performed the experiments and analyzed the samples; Pedro Yustos and Miguel Ladero developed the analytical procedures and analyzed the retrieved data; Miguel Ladero wrote the paper. All the authors read and approved the final manuscript.

Conflicts of Interest: The authors declare no conflict of interest. The founding sponsors had no role in the design of the study; in the collection, analyses, or interpretation of data; in the writing of the manuscript, and in the decision to publish the results.

References

1. Cherubini, F. The biorefinery concept: Using biomass instead of oil for producing energy and chemicals. *Energy Convers. Manag.* **2010**, *51*, 1412–1421. [CrossRef]
2. Ragauskas, A.J.; Williams, C.K.; Davison, B.H.; Britovsek, G.; Cairney, J.; Eckert, C.A.; Frederick, W.J.; Hallett, J.P.; Leak, D.J.; Liotta, C.L. The path forward for biofuels and biomaterials. *Science* **2006**, *311*, 484–489. [CrossRef] [PubMed]
3. Zhu, X.-G.; Long, S.P.; Ort, D.R. What is the maximum efficiency with which photosynthesis can convert solar energy into biomass? *Curr. Opin. Biotechnol.* **2008**, *19*, 153–159. [CrossRef] [PubMed]
4. Naik, S.N.; Goud, V.V.; Rout, P.K.; Dalai, A.K. Production of first and second generation biofuels: A comprehensive review. *Renew. Sust. Energy Rev.* **2010**, *14*, 578–597. [CrossRef]
5. Carriquiry, M.A.; Du, X.; Timilsina, G.R. Second generation biofuels: Economics and policies. *Energy Pol.* **2011**, *39*, 4222–4234. [CrossRef]
6. Pham, T.P.T.; Kaushik, R.; Parshetti, G.K.; Mahmood, R.; Balasubramanian, R. Food waste-to-energy conversion technologies: Current status and future directions. *Waste Manag.* **2015**, *38*, 399–408. [CrossRef] [PubMed]
7. Lal, R. World crop residues production and implications of its use as a biofuel. *Environ. Int.* **2005**, *31*, 575–584. [CrossRef] [PubMed]
8. FAOSTAT. Country Indicators, Data, Compare Data. 2014. Available online: http://www.fao.org/faostat/en/#home (accessed on 17 July 2017).
9. Erukainure, O.L.; Ebuehi, O.A.T.; Choudhary, M.I.; Mesaik, M.A.; Shukralla, A.; Muhammad, A.; Zaruwa, M.Z.; Elemo, G.N. Orange peel extracts: Chemical characterization, antioxidant, antioxidative burst, and phytotoxic activities. *J. Diet. Suppl.* **2016**, *13*, 585–594. [CrossRef] [PubMed]
10. Bampidis, V.A.; Robinson, P.H. Citrus by-products as ruminant feeds: A review. *Anim. Feed Sci. Technol.* **2006**, *128*, 175–217. [CrossRef]
11. Lessa, E.F.; Gularte, M.S.; Garcia, E.S.; Fajardo, A.R. Orange waste: A valuable carbohydrate source for the development of beads with enhanced adsorption properties for cationic dyes. *Carbohydr. Polym.* **2017**, *157*, 660–668. [CrossRef] [PubMed]
12. De Farias Silva, C.E.; da Silva Gonçalves, A.H.; de Souza Abud, A.K. Treatment of textile industry effluents using orange waste: A proposal to reduce color and chemical oxygen demand. *Water Sci. Technol.* **2016**, *74*, 994–1004. [CrossRef] [PubMed]

13. Miran, W.; Nawaz, M.; Jang, J.; Lee, D.S. Conversion of orange peel waste biomass to bioelectricity using a mediator-less microbial fuel cell. *Sci. Total Environ.* **2016**, *547*, 197–205. [CrossRef] [PubMed]
14. John, I.; Muthukumar, K.; Arunagiri, A. A review on the potential of citrus waste for D-Limonene, pectin, and bioethanol production. *Int. J. Green Energy* **2017**, *14*, 599–612. [CrossRef]
15. Putnik, P.; Kovačević, D.B.; Jambrak, A.R.; Barba, F.; Cravotto, G.; Binello, A.; Lorenzo, J.; Shpigelman, A. Innovative "Green" and Novel Strategies for the Extraction of Bioactive Added Value Compounds from Citrus Wastes—A Review. *Molecules* **2017**, *22*, 680. [CrossRef] [PubMed]
16. Rezzadori, K.; Benedetti, S.; Amante, E.R. Proposals for the residues recovery: Orange waste as raw material for new products. *Food Bioprod. Process.* **2012**, *90*, 606–614. [CrossRef]
17. Rafiq, S.; Kaul, R.; Sofi, S.A.; Bashir, N.; Nazir, F.; Nayik, G.A. Citrus peel as a source of functional ingredient: A review. *J. Saudi Soc. Agric. Sci.* **2016**. [CrossRef]
18. Sharma, K.; Mahato, N.; Cho, M.H.; Lee, Y.R. Converting citrus wastes into value-added products: Economic and environmently friendly approaches. *Nutrition* **2017**, *34*, 29–46. [CrossRef] [PubMed]
19. Ángel Siles López, J.; Li, Q.; Thompson, I.P. Biorefinery of waste orange peel. *Crit. Rev. Biotechnol.* **2010**, *30*, 63–69. [CrossRef] [PubMed]
20. Humbird, D.; Davis, R.; Tao, L.; Kinchin, C.; Hsu, D.; Aden, A.; Schoen, P.; Lukas, J.; Olthof, B.; Worley, M.; et al. *Process Design and Economics for Biochemical Conversion of Lignocellulosic Biomass to Ethanol: Dilute-Acid Pretreatment and Enzymatic Hydrolysis of Corn Stover*; National Renewable Energy Laboratory (NREL): Golden, CO, USA, 2011.
21. Mussatto, S.I. *Biomass Fractionation Technologies for a Lignocellulosic Feedstock Based Biorefinery*; Elsevier: Amsterdam, The Netherlands, 2016.
22. Rivas-Cantu, R.C.; Jones, K.D.; Mills, P.L. A citrus waste-based biorefinery as a source of renewable energy: Technical advances and analysis of engineering challenges. *Waste Manag. Res.* **2013**, *31*, 413–420. [CrossRef] [PubMed]
23. Santi, G.; Crognale, S.; D'Annibale, A.; Petruccioli, M.; Ruzzi, M.; Valentini, R.; Moresi, M. Orange peel pretreatment in a novel lab-scale direct steam-injection apparatus for ethanol production. *Biomass Bioenergy* **2014**, *61*, 146–156. [CrossRef]
24. Widmer, W.; Zhou, W.; Grohmann, K. Pretreatment effects on orange processing waste for making ethanol by simultaneous saccharification and fermentation. *Bioresour. Technol.* **2010**, *101*, 5242–5249. [CrossRef] [PubMed]
25. Wang, L.; Xu, H.; Yuan, F.; Fan, R.; Gao, Y. Preparation and physicochemical properties of soluble dietary fiber from orange peel assisted by steam explosion and dilute acid soaking. *Food Chem.* **2015**, *185*, 90–98. [CrossRef] [PubMed]
26. Merino, S.T.; Cherry, J. Progress and Challenges in Enzyme Development for Biomass Utilization. In *Biofuels*; Olsson, L., Ed.; Springer: Berlin/Heidelberg, Germany, 2007; pp. 95–120.
27. Klein-Marcuschamer, D.; Oleskowicz-Popiel, P.; Simmons, B.A.; Blanch, H.W. The challenge of enzyme cost in the production of lignocellulosic biofuels. *Biotechnol. Bioeng.* **2012**, *109*, 1083–1087. [CrossRef] [PubMed]
28. Fockink, D.H.; Urio, M.B.; Chiarello, L.M.; Sánchez, J.H.; Ramos, L.P. Principles and challenges involved in the enzymatic hydrolysis of cellulosic materials at high total solids. In *Green Fuels Technology: Biofuels*; Soccol, C.R., Brar, S.K., Faulds, C., Ramos, L.P., Eds.; Springer International Publishing: Cham, Switzerland, 2016; pp. 147–173.
29. Jung, Y.H.; Park, H.M.; Kim, D.H.; Yang, J.; Kim, K.H. Fed-Batch enzymatic saccharification of high solids pretreated lignocellulose for obtaining high titers and high yields of glucose. *Appl. Biochem. Biotechnol.* **2017**, *182*, 1108–1120. [CrossRef] [PubMed]
30. Wojtusik, M.; Zurita, M.; Villar, J.C.; Ladero, M.; Garcia-Ochoa, F. Influence of fluid dynamic conditions on enzymatic hydrolysis of lignocellulosic biomass: Effect of mass transfer rate. *Bioresour. Technol.* **2016**, *216*, 28–35. [CrossRef] [PubMed]
31. Corrêa, L.J.; Badino, A.C.; Cruz, A.J.G. Mixing design for enzymatic hydrolysis of sugarcane bagasse: Methodology for selection of impeller configuration. *Bioprocess Biosyst. Eng.* **2016**, *39*, 285–294. [CrossRef] [PubMed]
32. Satari, B.; Palhed, J.; Karimi, K.; Lundin, M.; Taherzadeh, M.J.; Zamani, A. Process optimization for citrus waste biorefinery via simultaneous pectin extraction and pretreatment. *BioResources* **2017**, *12*, 1706–1722. [CrossRef]

33. Hosseini, S.S.; Khodaiyan, F.; Yarmand, M.S. Optimization of microwave assisted extraction of pectin from sour orange peel and its physicochemical properties. *Carbohydr. Polym.* **2016**, *140*, 59–65. [CrossRef] [PubMed]

34. Sluiter, A.; Hames, B.; Ruiz, R.; Scarlata, C.; Sluiter, J.; Templeton, D.; Crocker, D. *Determination of Structural Carbohydrates and Lignin in Biomass*; National Renewable Energy Laboratory: Golden, CO, USA, 2008.

35. Sluiter, A.; Hames, B.; Ruiz, R.; Scarlata, C.; Sluiter, J.; Templeton, D. *Determination of Sugars, Byproducts, and Degradation Products in Liquid Fraction Process Samples*; National Renewable Energy Laboratory: Golden, CO, USA, 2006.

36. Sluiter, A.; Ruiz, R.; Scarlata, C.; Sluiter, J.; Templeton, D. *Determination of Extractives in Biomass*; National Renewable Energy Laboratory: Golden, CO, USA, 2008.

37. Dashtban, M.; Maki, M.; Leung, K.T.; Mao, C.; Qin, W. Cellulase activities in biomass conversion: Measurement methods and comparison. *Crit. Rev. Biotechnol.* **2010**, *30*, 302–309. [CrossRef] [PubMed]

38. Dalal, S.; Sharma, A.; Gupta, M.N. A multipurpose immobilized biocatalyst with pectinase, xylanase and cellulase activities. *Chem. Central J.* **2007**, *1*, 16. [CrossRef] [PubMed]

39. Bradford, M.M. A rapid and sensitive method for the quantitation of microgram quantities of protein utilizing the principle of Protein-Dye binding. *Anal. Biochem.* **1976**, *72*, 248–254. [CrossRef]

40. Wojtusik, M.; Rodríguez, A.; Ripoll, V.; Santos, V.E.; García, J.L.; García-Ochoa, F. 1,3-Propanediol production by *Klebsiella oxytoca* NRRL-B199 from glycerol. Medium composition and operational conditions. *Biotechnol. Rep.* **2015**, *6*, 100–107. [CrossRef] [PubMed]

41. Rao, R.S.; Kumar, C.G.; Prakasham, R.S.; Hobbs, P.J. The Taguchi methodology as a statistical tool for biotechnological applications: A critical appraisal. *Biotechnol. J.* **2008**, *3*, 510–523. [CrossRef] [PubMed]

42. Akhtar, N.; Goyal, D.; Goyal, A. Characterization of microwave-alkali-acid pre-treated rice straw for optimization of ethanol production via simultaneous saccharification and fermentation (SSF). *Energy Convers. Manag.* **2017**, *141*, 133–144. [CrossRef]

43. Kopelman, R. Fractal Reaction Kinetics. *Science* **1988**, *241*, 1620–1626. [CrossRef] [PubMed]

44. Wojtusik, M.; Zurita, M.; Villar, J.C.; Ladero, M.; Garcia-Ochoa, F. Enzymatic saccharification of acid pretreated corn stover: Empirical and fractal kinetic modelling. *Bioresour. Technol.* **2016**, *220*, 110–116. [CrossRef] [PubMed]

45. Fockink, D.H.; Urio, M.B.; Sánchez, J.H.; Ramos, L.P. Enzymatic hydrolysis of steam-treated sugarcane bagasse: Effect of enzyme loading and substrate total solids on its fractal kinetic modeling and rheological properties. *Energy Fuels* **2017**, *31*, 6211–6220. [CrossRef]

fermentation

MDPI

Article

Production of Fungal Biomass for Feed, Fatty Acids, and Glycerol by *Aspergillus oryzae* from Fat-Rich Dairy Substrates

Amir Mahboubi [1,2,*], Jorge A. Ferreira [1], Mohammad J. Taherzadeh [1] and Patrik R. Lennartsson [1]

[1] Swedish Centre for Resource Recovery, University of Borås, 50190 Borås, Sweden;
     Jorge.Ferreira@hb.se (J.A.F.); mohammad.taherzadeh@hb.se (M.J.T.); Patrik.Lennartsson@hb.se (P.R.L.)
[2] Flemish Institute for Technological Research, VITO NV, Boeretang 200, B-2400 Mol, Belgium
*   Correspondence: amir.mahboubi_soufiani@hb.se; Tel.: +46-33-4354612; Fax: +46-33-4354003

Received: 31 August 2017; Accepted: 19 September 2017; Published: 22 September 2017

Abstract: Dairy waste is a complex mixture of nutrients requiring an integrated strategy for valorization into various products. The present work adds insights into the conversion of fat-rich dairy products into biomass, glycerol, and fatty acids via submerged cultivation with edible filamentous fungi. The pH influenced fat degradation, where *Aspergillus oryzae* lipase was more active at neutral than acidic pH (17 g/L vs. 0.5 g/L of released glycerol); the same trend was found during cultivation in *crème fraiche* (12 g/L vs. 1.7 g/L of released glycerol). In addition to glycerol, as a result of fat degradation, up to 3.6 and 4.5 g/L of myristic and palmitic acid, respectively, were released during *A. oryzae* growth in cream. The fungus was also able to grow in media containing 16 g/L of lactic acid, a common contaminant of dairy waste, being beneficial to naturally increase the initial acidic pH and trigger fat degradation. Considering that lactose consumption is suppressed in fat-rich media, a two-stage cultivation for conversion of dairy waste is also proposed in this work. Such an approach would provide biomass for possibly feed or human consumption, fatty acids, and an effluent of low organic matter tackling environmental and social problems associated with the dairy sector.

Keywords: biomass for feed; dairy waste; edible filamentous fungi; fatty acids; glycerol

1. Introduction

Considering the important role of dairy products and related industries in the daily life and well-being of an increasing human population, it is also of environmental, social, and economic importance to have promising methods to treat the dairy waste generated. Therefore, efficient waste management routes need to be developed to cope with increasing amounts of dairy waste. One of those strategies includes biological treatment with edible filamentous fungi whose diversified enzymatic machinery enables the hydrolysis of complex nutrients for further assimilation and conversion to various value-added products [1,2]. Dairy substrates are potential sources of lactose, protein, and fat, where the latter varies between 3% in low-fat dairy substrates (e.g., milk) and 40%–50% in fat-rich substrates (e.g., cream and *crème fraiche*) [3]. Therefore, bringing together the metabolic diversity of filamentous fungi and the complex composition of dairy waste, there is a high potential for building biorefineries worldwide where dairy waste is biologically converted to various value-added products. This can have important environmental and social implications considering the amount of dairy waste generated, its environmental pollution potential, and biorefinery-related job creation.

Edible filamentous fungi such as *Aspergillus oryzae* and *Neurospora intermedia* have previously been investigated for valorization of dairy waste products including cheese-whey, milk, yoghurt, cream, and fermented cream (*crème fraiche*) and management routes and their output potential have been proposed [3]. Special emphasis was placed on lactose assimilation, production of biomass that can be used for feed applications, and fat degradation with concomitant release of glycerol. *N. intermedia* has been used for production of the Indonesian food *oncom*; however, it is a rather unexplored fungus from a biotechnological point of view [2]. *A. oryzae*, found worldwide but of higher incidence in tropical environments [4], is one of the most studied fungi at the industrial level where it contributes to the production of several products including a variety of organic acids and enzymes [5]. In addition, *A. oryzae* is traditionally used for production of various human food products (e.g., miso) and beverages (e.g., sake) in China and other East Asian countries [6]. Interestingly, *A. oryzae* was found to change its metabolic preference from lactose assimilation in low-fat dairy waste to fat degradation in high-fat dairy waste [3]. Therefore, fatty acids and glycerol, together with fungal biomass, would have a crucial role in the economic feasibility of bioconversion processes centered on dairy waste. Thus, further insights are needed regarding optimization of fat degradation by *A. oryzae* in dairy waste together with the characterization of the fatty acids released. Nearly 60% of milk fatty acids are saturated, where palmitic, stearic, and myristic acids dominate; polyunsaturated fatty acids such as linoleic acid only account for about 4%. Functional food monounsaturated fatty acids are also present in milk, for instance, oleic acid represents about 25% of milk fatty acid content [7–9].

The present study focuses on *A. oryzae* fat degradation capabilities in fat-rich substrates, namely cream and *crème fraiche* (pasteurized cream soured using lactic acid producing bacteria [10]). Special emphasis was given to the effect of pH and lactic acid on fat degradation and concomitant release of glycerol and fatty acids whose characterization was carried out. Bringing together all available insights on valorization of dairy substrates with edible filamentous fungi, an integrated bioconversion system for management of dairy waste is proposed.

2. Materials and Methods

2.1. Microorganism

Aspergillus oryzae var. oryzae CBS 819.72 (Centraalbureau voor Schimmelcultures, Utrecht, The Netherlands) was used throughout this study. The ascomycete was maintained in potato dextrose agar (PDA) plates containing 20 g/L glucose, 15 g/L agar, and 4 g/L potato extract. New plates were prepared on a two-week basis via inoculation with 100 µL of spore solution obtained by flooding a pre-grown plate with 20 mL sterile distilled water; disposable spreaders were used to bring the spores into solution and to spread the spore solution onto new plates. The inoculated plates were incubated at 30 °C for 3–5 days followed by storage at 4 °C until use for cultivation.

2.2. Dairy Substrates

Fresh cream (grädde, Falköping Mejeri, Falköping, Sweden) and *crème fraiche* (sour cream, ICA, Borås, Sweden) were purchased locally. Cream contained 48% (*w/w*) total solids composed of 40% fat, 3% carbohydrates, and 2% protein (Falköping Mejeri, Falköping, Sweden); *crème fraiche* contained 40% (*w/w*) solids composed of 34% fat, 3% carbohydrates, and 2% protein (ICA, Borås, Sweden). Both substrates were sterilized at 121 °C for 20 min followed by storage at 4 °C until use for cultivation. For cultivations in separated solid and liquid fractions of cream, it was centrifuged at 3000× *g* for 5 min and the solid fractions were diluted to the starting volume with distilled water. Both fractions were sterilized in 250 mL Erlenmeyer flasks containing 100 mL of medium at 121 °C for 20 min.

2.3. Cultivations

All cultivations were carried out using 250 mL Erlenmeyer flasks containing 100 mL of medium. The flasks were inoculated with 20 mL/L of solution containing 1.8×10^7 spores/L and incubated in a water bath shaking at 125 rpm at 35 °C for 4–6 days. Cultivations in non-sterile cream were carried out by transferring, under sterilized conditions, 100 mL cream into pre-sterilized empty 250 mL Erlenmeyer flasks. Cultivations in semi-synthetic medium were carried out as above using lactic acid (20 g/L) and glycerol (30 g/L) as single carbon sources together with salts and with or without supplementation of yeast extract according to Sues et al. [11]. Samples were withdrawn under sterile conditions during cultivation at regular intervals where the liquid fractions after centrifugation at $3000 \times g$ for 5 min were stored at $-20\,°C$ for further analyses. At the end of the cultivation, the biomass was harvested using a sieve (1 mm^2 pore size) and washed thoroughly with distilled water to remove extracellular medium residuals. All cultivations were carried out in duplicate.

2.4. Analyses

The harvested biomass was dried to constant weight in an oven at 70 °C and biomass yields are reported as g of biomass per liter of the dairy substrate.

Profiles of glycerol, lactic acid, lactose, glucose, other sugars, ethanol, and fatty acids were constructed based on high-performance liquid chromatography (HPLC) analysis. A hydrogen-ion based ion-exchange column (Aminex HPX-87H, Bio-Rad, Hercules, CA, USA) at 60 °C and 0.6 mL/min of 5 mM H_2SO_4 was used for analysis of all components except fatty acids. Myristic, palmitic, oleic, and stearic acids were analyzed using a C18 XBridge BEH Phenyl column (Waters Corporation, Milford, MA, USA) at 45 °C and 0.6 mL/min of acetonitrile-water (85:15). An ultraviolet (UV) absorbance detector (Waters 2487, Waters Corporation, Milford, MA, USA), operating at 210 nm wavelength, was used in series with a refractive index (RI) detector (Waters 2414).

Derivatization of the fatty acids in the experimental samples and respective standard mixtures was carried out as follows: 0.5 mL of the sample was dried under nitrogen stream followed by the addition of 0.5 mL of 20 mg bromoacetophenone/mL acetone and 0.5 mL of 25 mg triethylamine/mL acetone. The mixture was vortexed and then placed in a heating block at 100 °C for 15 min with the tubes open. After cooling, 0.75 mL of 10 mg acetic acid/mL acetone were added, heated at 100 °C for 5 min, and then dried under nitrogen stream. The volume of the samples was adjusted to 1 mL with acetonitrile-water (85:15), vortexed, centrifuged, and the supernatant was used for HPLC analysis.

The nitrogen content of solid and liquid fractions of cream was analyzed using the Kjeldahl method according to [3].

2.5. Statistical Analysis

All experiments performed in this study have been carried out in duplicate and the average values plus the error bars representing two standard deviations are illustrated on the graphs. The data acquired were statistically analyzed using MINITAB® 17 (Minitab Ltd., Coventry, UK). A general linear model with a confidence interval of 95% was applied for the analysis of variance (ANOVA). To have a better understanding of the extent of differences between results obtained, pairwise comparisons according to Tukey's test were performed.

3. Results and Discussion

Generally, dairy waste can be divided into fat-rich and low-fat substrates, where lactose contributes a great fraction of the chemical oxygen demand in the latter [12]. Understanding lactose consumption by *A. oryzae* and *N. intermedia* has previously been a subject of research [3]. Nonetheless, further investigation on fat degradation by edible filamentous fungi is needed in order to build a process that will lead to full conversion of the nutrients present in dairy waste to various value-added products. The present study focused on the influence of pH on fat degradation in order to give further insights regarding the optimal conditions for lipase production by *A. oryzae*, the lactic acid concentration as a result of bacterial growth, and the need for dairy waste sterilization before fungal cultivation. Furthermore, a characterization of the fatty acids released during fat degradation is presented together with a proposal for an integrated bioconversion process where all dairy waste is converted to various value-added products using edible filamentous fungi.

3.1. The Effect of pH on Fat Degradation by A. oryzae

Species belonging to the *Aspergillus* genus can grow at a wide range of pH values which together with temperature among other parameters, can influence enzyme production and activity [13]. Therefore, *A. oryzae* was grown in cream with pH initially adjusted from 4.3 to 7. As can be observed in Figure 1, pH played a crucial role on fat degradation, where the amount of glycerol released (as a result of fat degradation) increased from 0.5 g/L at acidic pH to 17.2 g/L at neutral pH. Differences in glycerol concentration were more evident up to pH 6 than within the pH interval 6–7 (Figure 1). Thus, at the examined conditions, production of lipase by *A. oryzae* was favored at neutral pH, which is in agreement with previous studies [14]. Neutral pH has also been observed to induce the highest lipase activity by *A. flavus*, while a pH of 6 was reported to induce the highest lipase activity in *A. niger* [15]. Therefore, pH values leading to the highest lipase activities vary naturally among *Aspergillus* species. The present research adds further insights into the biological conversion of dairy substrates by *A. oryzae*, since its lactase was found to have the highest activity within the pH range of 4–6 [3]. Furthermore, a temperature range of 30–37 °C is reported to induce maximal lipase activity by *Aspergillus* species [13–15]. Since the pH had not been kept constant during cultivation, a further experiment was carried out to establish a relationship between pH value and biomass production over time (Figure 2). The initial pH of cream after sterilization was 6.18 ± 0.11 and it decreased continuously to 5.36 ± 0.02 after 94 h and remained unchanged until the end of cultivation. As can be seen in Figure 2, as the pH reached 5.3, the biomass weight profile reached a plateau at ~5 g/L. This implies that low pH interferes with the growth of *A. oryzae* in cream media. It remains to be revealed how the decreasing pH influences the degradation of fat with consequent release of glycerol at longer cultivation times since a plateau has not been reached (Figure 1).

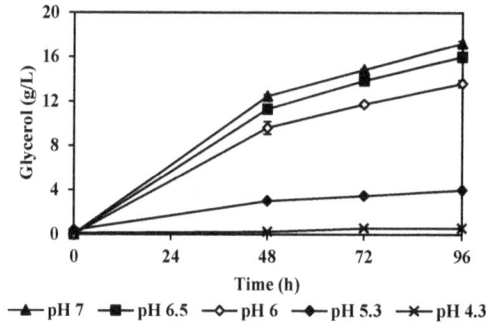

Figure 1. The effect of pH on fat degradation and glycerol release by *A. oryzae* in cream.

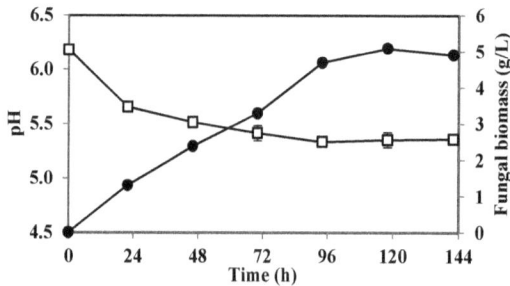

Figure 2. The trend of change in pH (□) and biomass (●) concentration in the medium containing cream and *A. oryzae*.

3.2. Fatty Acids Released by A. oryzae Fat Degradation

The contents of the fatty acids palmitic, myristic, stearic, and oleic acids were followed during cultivation in cream initially adjusted to pH 6. In complete agreement with the overtime release of glycerol, an increase in the release of the analyzed fatty acids was observed during cultivation (Figure 3). The final concentration of palmitic (4.56 ± 0.44 g/L) and myristic (3.68 ± 0.08 g/L) acids were considerably higher than that of stearic and oleic acids (less than 1 g/L of each). This difference in the release of fatty acids may be due to the type and activity of the lipase produced by *A. oryzae* [14,16]. Lipases are selective on the substrate; therefore, the final product depends on whether the initial substrates are more concentrated in tri-, di-, or mono-glycerides [17]. In addition, lipases are regioselective, with preference to act depending on the positioning of the fatty acids on the glycerol backbone. Svendsen [17] has reported that fungal lipase preferably acts on sn1 and sn3 positions on the triglyceride backbone. It is noteworthy that in milk fat triglyceride backbone, short chain fatty acids usually take sn3 and positions sn1 and sn2 are occupied by longer chain fatty acids [18].

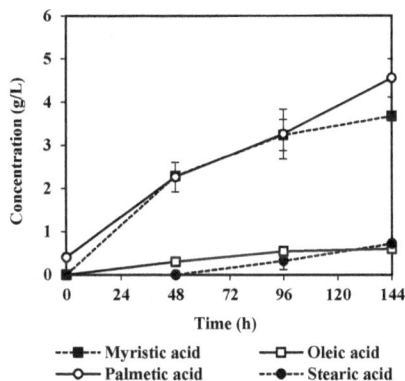

Figure 3. Changes in the concentration of main fatty acids in cream during the cultivation of *A. oryzae*.

However, when taking the lipid composition into account the detected glycerol concentration was more than 10 times higher than that of fatty acids. Thus, most likely the fungus is either metabolizing or storing the released fatty acids.

As reported by [3], in contradiction to low-fat dairy media, in cultivation of *A. oryzae* in fat-rich dairy cream medium there is a metabolic shift from lactose utilization to fat degradation. However, surprisingly, there seemed to be both very poor lactose consumption (lactose concentration remained constant throughout the cultivation [3]) and fat degradation in the case of *crème fraiche*, although both cream (40% fat) and *crème fraiche* (38% fat) are fat-rich dairy media. This activity hindrance may be due to the presence of about 11 g/L of lactic acid in the as-received *crème fraiche* that drops the initial pH to about 4.3, which according to previously obtained results is an unfavorable pH for fungal lipase activity. Therefore, the pH value effect on fat degradation was also investigated using *crème fraiche* as substrate (Figure 4). Expectedly, the increase in pH towards neutral range boosted fat degradation where glycerol release of up to 12 g/L at pH 6.5 was obtained.

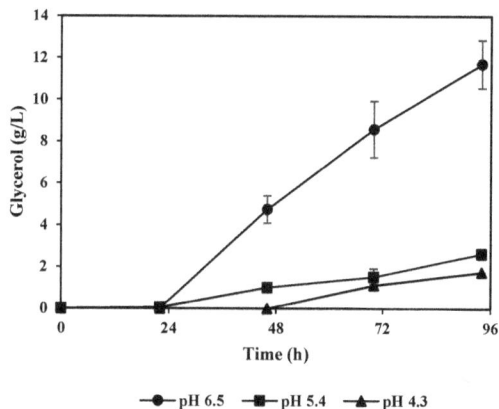

Figure 4. Glycerol yield in pH adjusted *crème fraiche* medium inoculated by *A. oryzae*.

3.3. Lactic Acid and Glycerol Consumption by A. oryzae

In non-sterile cream, as in the case of cream waste, the present bacterial consortium, mainly constituted by lactic acid bacteria, can grow on the lactose substrate, producing lactic acid as the main product. Lactic acid is partly behind the sour taste and rancid smell of expired dairy products. The lactic acid content of the medium as reported in the literature can act as fungicide [19,20], thus hindering the favorable metabolic activity of the fungal strain. On the other hand, as an increase in the concentration of lactic acid in the cultivation medium results in a drop of the pH, it may alter the trend of fat degradation by *A. oryzae*. Therefore, to study the effect of concomitant presence of *A. oryzae* and lactic acid bacteria in cream, the ability of *A. oryzae* to consume lactic acid as the sole carbon source was tested in semi-synthetic medium (Figure 5). Results from cultivation of *A. oryzae* in non-sterile cream can be found in a previous work [3]. As it can be observed, in the presence of a nitrogen source and 16 g/L of lactic acid the fungus was able to consume 37.5% of the acid within 96 h of cultivation. The degradation of lactic acid by different filamentous fungi including *A. oryzae* has been previously reported [3,21,22]. These results are promising when it comes to long cultivation periods, since if a stepwise process is applied, *A. oryzae* initially consumes lactic acid, and subsequently as the pH rises fat degradation will become dominant. This becomes even more critical when it comes to dairy waste, which contains considerable amounts of lactic acid bacteria that are constantly consuming lactose and producing lactic acid. In the presence of both the fungus and lactic acid bacteria, there seems to be no interference or interruption in the metabolic activity up to a certain initial lactic acid concentration, however, further metabolic behavior in this system is closely dependent on whether the lactic acid bacteria or fungus takes over the culture later in the cultivation process. This has also been proven in the work of de Vrese et al. [23]. Moreover, as *A. oryzae* has been reported to be a poor lactose consumer in fat-rich dairy media [3], parallel production of lactic acid by lactic acid bacteria that can be consumed by the fungus as a secondary substrate can be of interest from a process development point of view.

With the intensification of biodiesel production as a fuel replacement to diesel, the worldwide production of glycerol has concomitantly increased. Due to its surplus, the price of glycerol will have the tendency to decrease. Therefore, instead of considering glycerol as a final product, it is also considered in this study as a potential carbon source for *A. oryzae*. The fungus was grown in a semi-synthetic medium containing about 27 g/L glycerol as the sole carbon source together with magnesium, calcium, potassium, and ammonium salts, where the effect of absence or presence of yeast extract was investigated. As presented in Figure 5, when the medium was supplemented with yeast extract, 95% of the initial glycerol in the medium was consumed within 96 h of cultivation, while only 10% of the initial glycerol was consumed when yeast extract was absent. The results are in compliance with 7.09 ± 0.78 and 0.61 ± 0.13 g/L of biomass produced in cultivation media supplemented or not with yeast extract, respectively.

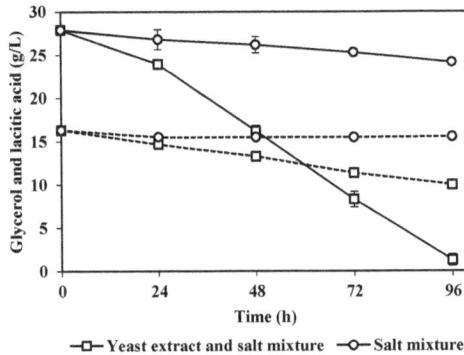

Figure 5. Utilization of lactic acid (dashed line) and glycerol (solid line) by *A. oryzae* in mediums supplemented with salt mixture (○) or salt and yeast extract (□).

3.4. Cultivation of A. oryzae in Different Fractions of Cream

Since the metabolic activity of *A. oryzae* alters in fat-rich dairy media, where lactose consumption is suppressed and lipid degradation encouraged, the ascomycete was cultivated in separated liquid and solid fractions of cream [3]. The results of cultivation of *A. oryzae* in cream fat and fat-free cream fractions are presented in Figure 6. The activity of the fungal strain in the fat-free portion of cream towards lactose utilization even at low and favorable pH 4.3 for *A. oryzae* lactase activity was low. Only 17%, 5%, and 5% of the initial lactose was assimilated by *A. oryzae* at pH 4.3, 5.3, and 6.3 by the end of cultivation, respectively (Figure 6A). Taking into account the C/N ratio of 4 obtained by the Kjeldahl analysis, the poor lactose consumption should not be related to lack of nitrogen but to other nutrients that remained in the solid fraction after centrifugation. According to statistical analysis of the results, lactose consumption at pH 4.3 was significantly higher (p value = 0.001) than that at 5.3 and 6.3, which relatively had no significant difference (p value = 0.126). These differences were also corroborated by the biomass yields obtained of 9.96 ± 1.90, 5.38 ± 0.42, and 2.48 ± 0.56 g/L after cultivation at pH 4.3, 5.3, and 6.3, respectively.

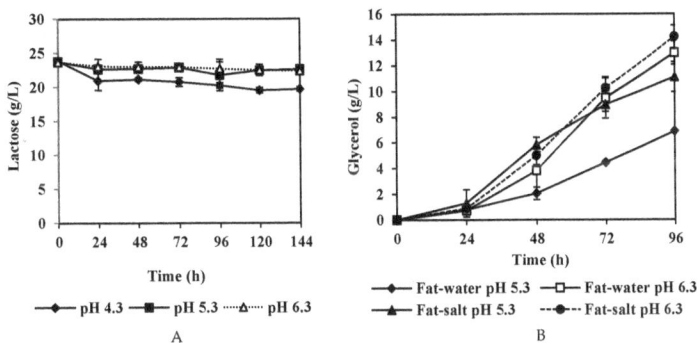

Figure 6. The changes in the (**A**) lactose content of the fat-free cream fraction inoculated with *A. oryzae* at different pH values and (**B**) glycerol concentration of the cream fat fraction inoculated with *A. oryzae* at different pH values and medium compositions.

Regarding the cultivation of *A. oryzae* in the fat-rich fraction of cream (Figure 6B) diluted with water or salt solution at different pH values, it can be observed that at pH 5.3 when the medium was provided with adequate salt mixture (including a nitrogen source in the form of ammonium), the concentration of glycerol released (11.11 ± 1.15 g/L) was significantly (p value = 0.005) higher than that when the cream fat was only diluted with water (6.88 ± 0.04 g/L). In comparison to salt supplemented medium at pH 5.3, no significant differences were observed regarding glycerol concentrations at pH 6.3 in cultures with water (12.98 ± 0.91 g/L) or salt solution (14.23 ± 0.88 g/L). Although the fungal biomass harvested from the salt supplemented cultivation medium at pH 6.3 (2.17 ± 0.12 g/L) was the highest of the four preparations, the intercellular lipid accumulated and released during oven drying of *A. oryzae* biomass [3] interferes with exact biomass measurements.

3.5. Proposed Integrated Bioconversion Unit of Dairy Substrates to Various Value-Added Products

The present study adds further insights into process requirements for full conversion of the nutrients in dairy waste to value-added products. *A. oryzae* has previously been found to change its metabolic preference, where fat degradation is preferred over lactose assimilation in fat-rich media and vice-versa. Such an expressive metabolic shift (based on the amount of glycerol released) has not been found for *Neurospora intermedia*, where lactose consumption assumes the highest relevance independently of the medium used [3]. Such performance differences call for an integrated process where both fungi are used in series (Figure 7) within a biorefinery-based concept for valorization of dairy waste. Overall, fat-rich dairy waste can be diluted with sour milk, cheese whey, or any other low-fat waste material to a fat content within 10–40% and used for cultivation with *A. oryzae* in a first bioreactor under aerobic conditions. During this first stage, *A. oryzae* will degrade fat and likely proteins, releasing fatty acids and glycerol. The medium is filtrated into a second aerobic bioreactor where *N. intermedia* receives a stream rich in lactose, lactic acid, glycerol, and fatty acids and produces biomass that can be easily separated by sieving and sold possibly for animal feed or human food applications. Despite the edible character of *A. oryzae*, further studies need to be performed within the overall process development in order to evaluate its suitability for animal feed or human consumption, since the growth substrate will be very heterogeneous. Depending on the conditions applied, the final liquid fraction should be very low in organic matter, which eases water recycling, and with fatty acids that can also have feed or human food applications. Further studies are needed in order to investigate the process performance at reactor scale and the influence of, for example, oxygen. Altogether, filamentous fungi can be used to tackle a social and environmental problem of increasing amounts of dairy waste, via building an integrated bioconversion process that can supply several different products into various industrial sectors.

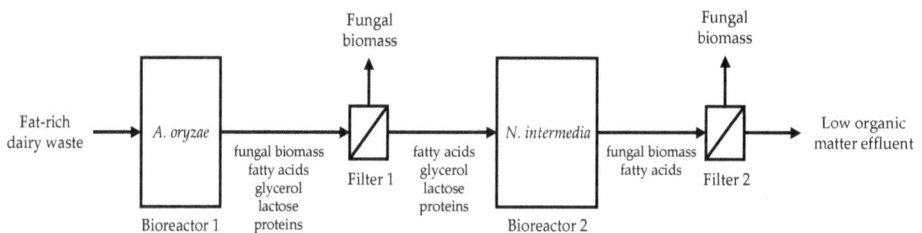

Figure 7. Schematic of the proposed integrated bioconversion process stages for fungal bioconversion of fat-rich dairy substrates to value-added products.

4. Conclusions

The present study adds important insights on the valorization of fat-rich substrates. The pH was found to have a critical effect on fat degradation by *A. oryzae* during growth in cream and

Fermentation **2017**, *3*, 48

fermented crème (*crème fraiche*). The concentration of released glycerol was 0.5 and 17 g/L during fungal cultivation in cream at acidic and neutral pH, respectively; a range of 1.7–12 g/L was observed when changing from acidic to neutral pH in *crème fraiche*. It was also found that a high concentration of lactic acid, due to contamination, will not have a negative impact on fat degradation by *A. oryzae*. Due to suppression of lactose consumption in fat-rich media, a two-stage cultivation using *A. oryzae* as a fat degrader and *Neurospora intermedia* as a lactose consumer is hypothesized to be the right strategy to build a biorefinery resembling process around dairy waste for supply of biomass for feed or human consumption, fatty acids, and low organic matter-containing effluent of easy disposal, or for easy water recycling.

Acknowledgments: This work was financially supported by Swedish Research Council Formas.

Author Contributions: Amir Mahboubi carried out the experimental work and part of the writing; Jorge A. Ferreira was responsible for part of the writing and contributed to the discussion during the research work; Mohammad J. Taherzadeh and Patrik R. Lennartsson developed the idea and contributed to the discussion during the research and the revision of the manuscript.

Conflicts of Interest: The authors declare no conflicts of interest.

References

1. Ferreira, J.A.; Lennartsson, P.R.; Edebo, L.; Taherzadeh, M.J. Zygomycetes-based biorefinery: Present status and future prospects. *Bioresour. Technol.* **2013**, *135*, 523–532. [CrossRef] [PubMed]
2. Ferreira, J.A.; Mahboubi, A.; Lennartsson, P.R.; Taherzadeh, M.J. Waste biorefineries using filamentous ascomycetes fungi: Present status and future prospects. *Bioresour. Technol.* **2016**, *215*, 334–345. [CrossRef] [PubMed]
3. Mahboubi, A.; Ferreira, J.A.; Taherzadeh, M.J.; Lennartsson, P.R. Value-added products from dairy waste using edible fungi. *Waste Manag.* **2017**, *59*, 518–525. [CrossRef] [PubMed]
4. Barbesgaard, P.; Heldt-Hansen, H.P.; Diderichsen, B. On the safety of *Aspergillus oryzae*: A review. *Appl. Microbiol. Biotechnol.* **1992**, *36*, 569–572. [CrossRef] [PubMed]
5. Gibbs, P.A. Growth of filamentous fungi in submerged culture: Problems and possible solutions. *Crit. Rev. Biotechnol.* **2000**, *20*, 17–48. [CrossRef] [PubMed]
6. Bentley, R. From miso, sake and shoyu to cosmetics: A century of science for kojic acid. *Nat. Prod. Rep.* **2006**, *23*, 1046–1062. [CrossRef] [PubMed]
7. Bauman, D.E.; Lock, A. Milk fatty acid composition: Challenges and opportunities related to human health. In Proceedings of the XXVI World Buiatrics Congress, Santiago, Chile, 14–18 November 2010; Cornell University: New York, NY, USA, 2010; pp. 278–289.
8. Nettleton, J. Introduction to Fatty Acids. In *Omega-3 Fatty Acids and Health*; Springer: Berlin, Germany, 1995; 63p.
9. Yilmaz-Ersan, L. Fatty acid composition of cream fermented by probiotic bacteria. *Mljekarstvo* **2013**, *63*, 132–139.
10. Meunier-Goddik, L. Sour Cream and Creme Fraiche. In *Handbook of Food and Beverage Fermentation Technology*; CRC Press: Boca Raton, FL, USA, 2004.
11. Sues, A.; Millati, R.; Edebo, L.; Taherzadeh, M.J. Ethanol production from hexoses, pentoses, and dilute-acid hydrolyzate by *Mucor indicus*. *FEMS Yeast Res.* **2005**, *5*, 669–676. [CrossRef] [PubMed]
12. Guimarães, P.M.R.; Teixeira, J.A.; Domingues, L. Fermentation of lactose to bio-ethanol by yeasts as part of integrated solutions for the valorisation of cheese whey. *Biotechnol. Adv.* **2010**, *28*, 375–384. [CrossRef] [PubMed]
13. Xia, J.L.; Huang, B.; Nie, Z.Y.; Wang, W. Production and characterization of alkaline extracellular lipase from newly isolated strain *Aspergillus awamori* HB-03. *J. Cent. South Univ. Technol.* **2011**, *18*, 1425–1433. [CrossRef]
14. Toida, J.; Kondoh, K.; Fukuzawa, M.; Ohnishi, K.; Sekiguchi, J. Purification and characterization of a lipase from *Aspergillus oryzae*. *Biosci. Biotechnol. Biochem.* **1995**, *59*, 1199–1203. [CrossRef] [PubMed]

Fermentation **2017**, *3*, 48

15. Colla, L.M.; Ficanha, A.M.M.; Rizzardi, J.; Bertolin, T.E.; Reinehr, C.O.; Costa, J.A.V. Production and characterization of lipases by two new isolates of Aspergillus through solid-state and submerged fermentation. *BioMed Res. Int.* **2015**. [CrossRef] [PubMed]

16. Ohnishi, K.; Yoshida, Y.; Toita, J.; Sekiguchi, J. Purification and characterization of a novel lipolytic enzyme from *Aspergillus oryzae. J. Ferment. Bioeng.* **1994**, *78*, 413–419. [CrossRef]

17. Svendsen, A. Lipase protein engineering. Biochim. Biophys. Acta (BBA) Protein Struct. *Mol. Enzymol.* **2000**, *1543*, 223–238. [CrossRef]

18. Fox, P.F.; McSweeney, P.L.H. *Dairy Chemistry and Biochemistry*; Springer: Berlin, Germany, 1998.

19. Gerez, C.L.; Torres, M.J.; Font de Valdez, G.; Rollán, G. Control of spoilage fungi by lactic acid bacteria. *Biol. Control* **2013**, *64*, 231–237. [CrossRef]

20. Muhialdin, B.J.; Hassan, Z. Screening of Lactic Acid Bacteria for Antifungal Activity against *Aspergillus oryzae. Am. J. Appl. Sci.* **2011**, *8*, 447–451. [CrossRef]

21. Torres, A.; Li, S.M.; Roussos, S.; Vert, M. Screening of microorganisms for biodegradation of poly (lactic-acid) and lactic acid-containing polymers. *Appl. Environ. Microbiol.* **1996**, *62*, 2393–2397. [PubMed]

22. Torres, A.; Li, S.M.; Roussos, S.; Vert, M. Degradation of L- and DL-lactic acid oligomers in the presence ofFusarium moniliforme andPseudomonas putida. *J. Environ. Polym. Degrad.* **1996**, *4*, 213–223. [CrossRef]

23. De Vrese, M.; Laue, C.; Offick, B.; Soeth, E.; Repenning, F.; Thoß, A.; Schrezenmeir, J. A combination of acid lactase from *Aspergillus oryzae* and yogurt bacteria improves lactose digestion in lactose maldigesters synergistically: A randomized, controlled, double-blind cross-over trial. *Clin. Nutr.* **2014**, *34*, 394–399. [CrossRef] [PubMed]

fermentation

MDPI

Review

Utilization of Volatile Fatty Acids from Microalgae for the Production of High Added Value Compounds

Angelina Chalima [1], Laura Oliver [2], Laura Fernández de Castro [2], Anthi Karnaouri [1], Thomas Dietrich [2] and Evangelos Topakas [1,3,*]

[1] IndBioCat group, Biotechnology Laboratory, School of Chemical Engineering, National Technical University of Athens, 5 Iroon Polytechniou Str., Zografou Campus,15780 Athens, Greece; achalima@chemeng.ntua.gr (A.C.); akarnaouri@chemeng.ntua.gr (A.K.)
[2] Food & Health Group, Health Division, Tecnalia Research & Innovation, Parque Tecnológico de Álava, Leonardo Da Vinci, 11, E-01510 Miñano-Álava, Spain; laura.oliver@tecnalia.com (L.O.); laura.fernandezdecastro@tecnalia.com (L.F.d.C.); thomas.dietrich@tecnalia.com (T.D.)
[3] Biochemical and Chemical Process Engineering, Division of Sustainable Process Engineering, Department of Civil Environmental and Natural Resources Engineering, Luleå University of Technology, SE-97187 Luleå, Sweden
* Correspondence: vtopakas@chemeng.ntua.gr; Tel.: +30-210-772-3264

Received: 9 September 2017; Accepted: 9 October 2017; Published: 15 October 2017

Abstract: Volatile Fatty Acids (VFA) are small organic compounds that have attracted much attention lately, due to their use as a carbon source for microorganisms involved in the production of bioactive compounds, biodegradable materials and energy. Low cost production of VFA from different types of waste streams can occur via dark fermentation, offering a promising approach for the production of biofuels and biochemicals with simultaneous reduction of waste volume. VFA can be subsequently utilized in fermentation processes and efficiently transformed into bioactive compounds that can be used in the food and nutraceutical industry for the development of functional foods with scientifically sustained claims. Microalgae are oleaginous microorganisms that are able to grow in heterotrophic cultures supported by VFA as a carbon source and accumulate high amounts of valuable products, such as omega-3 fatty acids and exopolysaccharides. This article reviews the different types of waste streams in concert with their potential to produce VFA, the possible factors that affect the VFA production process and the utilization of the resulting VFA in microalgae fermentation processes. The biology of VFA utilization, the potential products and the downstream processes are discussed in detail.

Keywords: volatile fatty acids; waste valorization; microalgae; bioactive compounds; omega-3 fatty acids

1. Introduction

The term Volatile Fatty Acids (VFA) is applied to short-chain fatty acids, usually consisted of two to six carbon atoms, such as acetic, butyric or propionic acid [1]. VFA are organic chemicals with various applications; they can be provided as carbon sources to microorganisms that produce useful metabolites or remove organic pollutants from waste water, they are utilized for electricity or hydrogen generation and they can serve as starting materials for the synthesis of long-chain fatty acids and polyhydroxyalkanoates (PHAs) for packaging applications [2].

VFA can be easily produced by all types of biomass (terrestrial, marine and aquatic) and within the frame of VFA platform, they can be used for the production of biofuels and biochemicals, offering a solution for efficient waste management. Valorization of VFA can be achieved via their efficient transformation into value-added products. Bioactive compounds such as omega-3 fatty acids can be obtained through fermentation processes that use VFA as carbon source for microorganisms.

According to recent research, some microalgae, able to grow heterotrophically, are capable of utilizing dissolved carboxylic acids such as VFA as carbon source [3]. Microalgae use short-chain fatty acids to produce metabolites, such as long chain unsaturated fatty acids (e.g., omega-3 fatty acids or arachidonic acid) or carotenoids. In this case, the first step of breaking down the carbon sources to simple sugars is eliminated, since VFA provide microalgae with a carbon chain backbone ready to be elongated to polyunsaturated fatty acids (PUFAs) [4].

This ability of certain microalgal species is opening a new perspective in the industrial production of high added-value products, while using as raw material undesirable persistent pollutants. The main metabolites of microalgae are lipids able to be esterified to produce biodiesel, long-chain polyunsaturated fatty acids and pigments. Several pigments and PUFAs, such as omega-3 fatty acids, are valuable products, with great importance for the food and pharmaceutical market. Therefore, the development of appropriate fermentation techniques that utilize VFA not only poses a solution to the elimination of these pollutants, but also offers an affordable production process of useful metabolites.

Due to the rising awareness among consumers about healthy and balanced diet, the market for omega-3 fatty acids especially eicosapentaenoic acid (EPA) and docosahexaenoic acid (DHA) is supposed to grow at a Compound Annual Growth Rate (CAGR) of 14.9% from 2016 to 2022 reaching a value of $6955 million by 2022 [5]. In general, fish oil or fish is a good source for DHA and EPA. Nevertheless, as the demand for omega-3 fatty acids is continuously rising and more and more fish is coming from aquaculture, it comes inevitably that traditional finite marine ingredients (e.g., fish oil) are replaced by terrestrial oils leading to a reduction of EPA and DHA and thereby compromising the nutritional value of the final fish product. Between 2006 and 2015, the amount of EPA and DHA requiring a double portion size in order to satisfy recommended levels [6]. Therefore, new sources (e.g., heterotrophic microalgae) and innovative production processes not relying on agricultural area and food grade carbon sources are needed to supply this gap.

2. Production of VFA from Waste Streams

2.1. VFA Production through Dark Fermentation

The synthesis of VFA is usually carried out through petrochemical processes that have a hazardous environmental impact [7]. On the other hand, VFA can be produced from food wastes, sludge, and a variety of biodegradable organic wastes via dark fermentation process, making the process more efficient and profitable from economic and environmental point of view. Dark fermentation initially follows the same pattern as the classical anaerobic digestion (hydrolysis, acidogenesis, acetogenesis) with the difference that the final step of methanogenesis is inhibited by various methods [8]. Hydrolysis, the first and usually the slowest step part of anaerobic digestion, includes the enzymatic breakdown of long polymeric substances to simpler organic monomers such as sugars, amino acids and fatty acids. These monomers are subsequently fermented by acidogenic populations in the acidogenesis step, leading to production of VFA (mainly acetate), together with an air mixture of hydrogen and carbon dioxide. Acetogenesis represents the stage where the breakdown of VFA to hydrogen and acetate occurs [9]. In order to utilize dark fermentation to produce VFA, the consumption of fatty acids by methanogens must be prevented. Therefore, in the above process, the last step of methanogenesis is inhibited, usually either by thermal pretreatment of the waste stream to destroy the methanogenic populations, or by maintaining the pH of the mixture at high values (above 9) that do not allow the growth of the specific bacteria [10,11]. Another solution is the addition of a methanogen inhibitor, such as iodoform [12,13]. Due to methanogenesis inhibition, dark fermentation usually leads to hydrogen and VFA accumulation, especially high carbon fatty acids such as caproate, production of ethanol and accumulation of lactate [14,15].

As already mentioned, the slowest and, therefore, rate-limiting step of dark fermentation is the hydrolysis of solids in the waste stream [11]. This is related to the overall recalcitrance of the biomass structure due to different factors that render the substrate unavailable to hydrolysis, as well as the presence of compounds, such as phenolics and the low C/N ratio together with high levels of trace elements that inhibit the degradation [16,17]. In order to increase VFA production, the hydrolysis rate and/or efficiency should be enhanced. The biodegradability of the waste materials can be improved by the application of an initial pretreatment step that will convert raw materials to a form amenable to enzymatic and microbial degradation. As a result, the vast majority of dark fermentation applications usually involve the use of a suitable pretreatment process of the waste stream prior to biological treatment. Hydrolysis can be enhanced either by thermal, ultrasonic, microwave or chemical pretreatment with acid, alkali, surfactants, ozone or other reactive substances [18–20]. These methods damage cell walls, releasing intracellular substances and/or enhance the solubilization of extracellular matter. An alternative way to boost the hydrolysis step is to apply an enzymatic treatment with commercial enzyme mixtures, where specific activities targeting particular waste components can be used, while different parameters like enzyme dosage and hydrolysis time can be controlled towards maximum monomers yields [21]. Hee Jun Kim et al., for example, used a mixture of the commercial enzymes Viscozyme, Flavourzyme and Palatase (carbohydrase, protease, lipase) to pretreat raw food waste in order to maximize VFA production during dark fermentation. The new production rate of VFA was 3.3 times higher after the enzymatic treatment [22]. Another commercial enzyme cocktail that has been applied to enhance the solubilization of pulp and paper mill secondary sludge is Accellerase 1500, known for its applications on cellulose hydrolysis [23]. To the best of our knowledge a certain enzyme available for every type of waste stream has not yet been identified. However, Q. Yang et al. have concluded that a treatment of sludge with amylases is more efficient than one with proteases [24]. Since an enzymatic treatment can increase the cost of the whole process, it is suggested not to rely solely on enzymes to enhance waste hydrolysis. Usually a combination of the previously listed techniques is preferable.

An established method that includes the production of VFA through dark fermentation is the MixAlco™ process. The MixAlco™ process is used to convert biomass into carboxylate salts through dark fermentation by a mixed-culture of microorganisms. In a second step of the process the salts are converted to ketones and fuels. The interesting part of this technique is that it utilizes countercurrent fermentation through a number of fermenters, thus allowing the feed with the lowest VFA concentration to mix with the most fermented biomass and vice versa [25]. The advantage of this is based on the fact that the products of the fermentation process, namely VFA, can act as inhibitors to the growth of acidogens, resulting in slow production rates [26].

2.2. Available Waste Streams for VFA Production and VFA Yields

All types of biomass can be used as substrates for biogas production as long as they contain carbohydrates, proteins, fats, cellulose, and hemicelluloses as main components. Waste streams that are deemed appropriate for VFA production should have a high organic load that can support the growth of acidogenic bacteria. Amongst the most popular waste streams for dark fermentation are various types of sludge, wastewaters from paper or agricultural industries and food waste (Table 1). Some of the streams that are being used, like waste activated sludge, already include the appropriate bacteria population, whilst others have to be inoculated with an external bacterial consortium.

In many applications, the streams that are destined for VFA production are a mixture of different waste streams e.g., agricultural waste and animal manure [13] or waste activated sludge (WAS) and corn stover [27]. The purpose behind mixing different waste streams is the development of a substrate with a balanced carbohydrate: protein ratio, which in turn answers for a balanced C/N ratio. According to various experiments [27,28] it has been found that a higher C/N ratio (around 20/1–30/1) can be beneficial to the microorganisms responsible for acidogenesis. Furthermore, it is believed that the carbohydrate amount can have a synergistic effect with WAS. Therefore, the addition of a rich in carbohydrate-waste stream to the original substrate can lead to enhanced VFA productivity [29]. However, it does not follow that a rich in protein-waste stream would not be able to produce VFA through dark fermentation. Pessiot et al. have managed to produce a considerable amount of VFA from slaughterhouse waste [30]. Due to the high amount of nitrogen available in the specific waste stream, the production of N_2 emissions and nitrogen ions, that buffered successfully the mixture, was also observed, together with VFA production. As it is, the use of waste rich in nitrogen seems to offer a different advantage, which is the ability to maintain a high pH value during the fermentation, without the need of a buffer feed.

The liquid product stream of dark fermentation processes consists of a mixture of VFA with different compositions. In the vast majority, the dominant acid in these mixtures is acetic acid, followed by propionic acid (Table 1). Acetic acid seems to have a higher value as a product, since it is more easily consumed by several microorganisms [31]. However, since it is classified as a hazardous pollutant for the environment, its production through dark fermentation needs to be justified by its utilization as a raw material for the production of high added-value products. The composition of the acid fraction of the liquid product depends on various variables, like the type of waste stream used and the fermentation conditions (pH, fermentation time, etc.) [1,28]. The pH value for example plays an important role in the synthesis of different VFA. It has been reported in the literature that, during the fermentation of a synthetic gelatin wastewater, pH values between 4.0–5.0 favors the production of propionic acid, while a pH around 6.0–7.0 favors the production of acetic and butyric acid, instead [32]. Chen et al. also noticed that, during the fermentation of a mixture of primary and waste activated sludge, the proportion of acetate increased when the pH increased from 7.9 to 8.9 [33].

Regarding the composition of waste, it has been proposed that a higher C/N ratio, under the appropriate pH conditions, favors the production of propionate. Propionic acid is supposed to enhance the efficiency of biological phosphorus removal from wastewater, when VFA are used as a carbon source for nutrient removal [28]. For that reason, Feng et al. mixed WAS with boiled rice in order to produce a rich in carbohydrates substrate suitable for VFA production.

2.3. VFA Platform for Fuels and Chemicals

The development of a platform that enables the production of a mixture of VFA from bio-waste by dark fermentation, thus consuming the organic load of this waste, can offer instant environmental relief. When compared to other biorefinery platforms, such as syngas (thermochemical), sugar and biogas, the VFA platform offers unique advantages [34]. This platform enables the use of all types of biomass and waste streams, leading to high productivity levels (high VFA yields) with low production costs, while enabling possible hydrogen co-production. Moreover, it has lower CO_2 emission than a typical sugar platform.

In order to ensure that VFA utilization processes will not release to the environment pollutants such as acetic acid, it is necessary to convert VFA to high added-value substances in a sustainable manner. Fermentation techniques have gained immense importance due to their economic and environmental advantages. Establishing a VFA platform and incorporating fermentative production routes towards value-added compounds, comprise a key step for the implementation of sustainable value chains for the use of biomass wastes. In this context, the European funded research project VOLATILE (Grant Agreement No. 720777) offers a great opportunity to transform municipal solid and sludgy biowaste into VFA to be used as carbon sources for value-added fermentation approaches to obtain omega-3 fatty acids, the biopolymer PHA as well as single cell oil.

Table 1. Different waste streams used for Volatile Fatty Acids (VFA) production.

Waste Stream	Process	Fermentation Key Conditions [1]	VFA Yield [2]	VFA Composition (mol)	Ref.
Pulp and paper mill effluent	Continuous	pH = 6, RT = 24 h, T = 37 °C	0.75 g COD/g COD	33Ac:61Pr:2Bu	[35]
Dairy whey effluent	Continuous	pH = 6, RT = 95 h, T = 37 °C	0.93 g COD/g COD	31Ac:41Pr:10Bu	[35]
Swine manure	Batch (2d)/ Semi-continuous (10d)	T = 38 ± 1 °C, pH = 5.3–5.6, HRT = 3 d	2002.25 mg $L^{-1}d^{-1}$		[10]
Waste Activated Sludge (WAS)	Batch	T = 35 ± 1°C, pH = 10	129.21 mg/g VS		[36]
Mixture of primary sludge (PS)/ WAS (*w/w*: 1:1)	Semi-continuous	T = 21± 1 °C, SRT = 6 d	118.4 ± 5.8 mg COD/g VSS	10Ac:7Pr:4Bu:5Va	[33]
Excess Sludge (ES)	Batch	T = 28 °C, pH = 10	302.4 mg COD/g VSS	53Ac:24Pr:9Bu: 9Iso-bu:9Iso-va	[37]
80% lime-treated sugarcane bagasse/20% chicken manure	Continuous	T = 55 °C, pH = 6.95–7.05, LRT = 19.1 d, VSLR = 2.07 g/(L·day)	0.55 g TA/g VS digested	91Ac:2Pr:7Bu:0.6Va	[13]
Slaughterhouse by-products	Fed-batch	T = 38 °C, $pH_{initial}$ = 6.8	0.38 ± 0.04 g VFA/g DM consumed	63Ac:43Pr:59Bu: 13Iso-Bu:24Va	[30]
80% shredded office copier paper/20% chicken manure	Semi-continuous	T = 40 °C, pH = 5.3-6.6, LRT = 32.6 d	0.159 g acid/g NAVS fed[3]	36Ac:37Pr:7Bu: 14Va:6Hep	[25]
WAS/ rice	Batch	T = 21±1 °C, pH = 8, fer. Time = 8 d	520.1±24.4 mg COD/g VSS	35Ac:50Pr	[28]
Alkaline, thermal pretreated WAS/ ABS	Batch/Semi-continuous	T = 35 ± 2 °C, HRT=10 d	712 ± 49 mg COD/g VSS	46 ± 0.9% Ac	[18]
Pretreated WAS/ pretreated corn stover (50:50)	Batch	T = 35 ± 1 °C, $pH_{initial}$ = 10	11939 mg COD/L	52Ac:22Pr:10n-Bu:8iso-Va	[27]
WAS pretreated with rhamnolipids (0.04 g/g TSS)	Batch	T = 35 ± 1 °C, fer. time = 4 d	3840 mg COD/L	Ac (25 ± 0.6%): Pr (20 ± 0.1%): n-HBu (16 ± 0.1%)	[20]

[1] SRT: Sludge Retention Time, HRT: Hydraulic Retention Time, LRT: Liquid Retention time, VSLR: volatile solids loading rate. [2] COD/sCOD: (Soluble) Chemical Oxygen Demand, VS/VSS: volatile (suspended) solids, TA: total acids, DM: dry matter, NAVS: non-acid volatile solid. [3] corresponds to total acid produced, whether in the liquid product or the solid waste. T: temperature; d: day.

3. Microalgae Potential in VFA Valorization

Microalgae are microorganisms that can grow both phototrophically and heterotrophically. When under dark, some microalgae strains can consume a carbon source and grow, while producing useful metabolites, such us lipids (e.g., omega-3 fatty acids) and pigments [3]. It has also been found that certain microalgae species would grow on different carbon sources, such as glucose, glycerol, ethanol and volatile organic acids. Apart from pure organic acids, VFA derived from dark fermentation processes have also been used as a carbon source for the cultivation of certain microalgae strains, like *Chlorella vulgaris* and *Auxenochlorella prototothecoides* [3,38]. Therefore, microalgae appear as a mean of bioconversion of VFA from fermented biowaste to high added-value products.

3.1. Biology of VFA Utilization;

Microalgae are ubiquitous organisms that are extremely diverse and heterogeneous from evolutionary and ecological point of view. They constitute prokaryotic cyanobacteria and eukaryotic protists and therefore exhibit variation in their nutritional requirements, as well as metabolite production. Commonly, microalgae are considered photoautotrophic organisms; however, several species use heterotrophic metabolism [39]. The interest in heterotrophic microalgae production is driven by the fact that it addresses many of the problems that have hampered the successful widespread implementation of photosynthetic microalgae cultivation systems [40].

Heterotrophic growth using carboxylic acids, such as acetic, citric, fumaric, glycolic, lactic, malic, pyruvic and succinic, as a substrate has been demonstrated for microalgae species [41]. Acetic acid is an intermediate of anaerobic digestion. It often accumulates in dark fermentation processes and can be easily converted by microalgae into acetyl-CoA, the main precursor for lipid synthesis [42]. Eukaryotic microorganisms are able to assimilate acetate via a monocarboxylic/proton transporter protein that aids transport of monocarboxylic molecules across the membrane [3] (Figure 1). The transferred acetate is used from the acetyl-CoA synthetase to acetylate coenzyme A using a single ATP molecule [39].

The metabolic oxidation of acetyl-CoA is using two pathways. In the glyoxysomes, the glyoxylate cycle transforms the acetyl-CoA into malate. The second pathway is converting acetyl-CoA into citrate using the tricarboxylic acid (TCA) cycle in the mitochondria, thereby providing carbon skeletons, energy as ATP, and energy for reduction as NADH [4,43]. The glyoxylate cycle is a pathway that allows the synthesis of four carbon metabolites from acetyl-CoA and it is similar to the Krebs cycle [43]. Isocitrate lyase and malate synthetase are the two specific enzymes of the glyoxylate cycle. Both enzymes are induced when cells are transferred to media containing acetate [3]. However, for many microorganisms, acetate could be toxic at high concentrations. Therefore, acetate concentration at low level should be used in fed-batch cultures or under pH-auxostat cultivation conditions keeping pH value constant [3].

In order to induce fatty acid synthesis nitrogen starvation may be used. It could be shown that in this case the glyoxylate pathway, as well as activity of acetyl-CoA synthetase was down regulated [44,45]. Thus, *Chlamydomonas reinhardtii* is using a second pathway via acetate kinase and phosphate acetyltransferase for acetate assimilation, thereby maintaining basic cellular functions and providing energy [45]. Excess carbon can be directly used for fatty acid biosynthesis [44]. Another important aspect during N-deprivation is the reduced activity of isocitrate dehydrogenase resulting in an increase of citrate levels. Citrate can be converted into oxaloacetate and acetyl-CoA via ATP citrate lyase providing further acetyl-CoA for fatty acid biosynthesis [45].

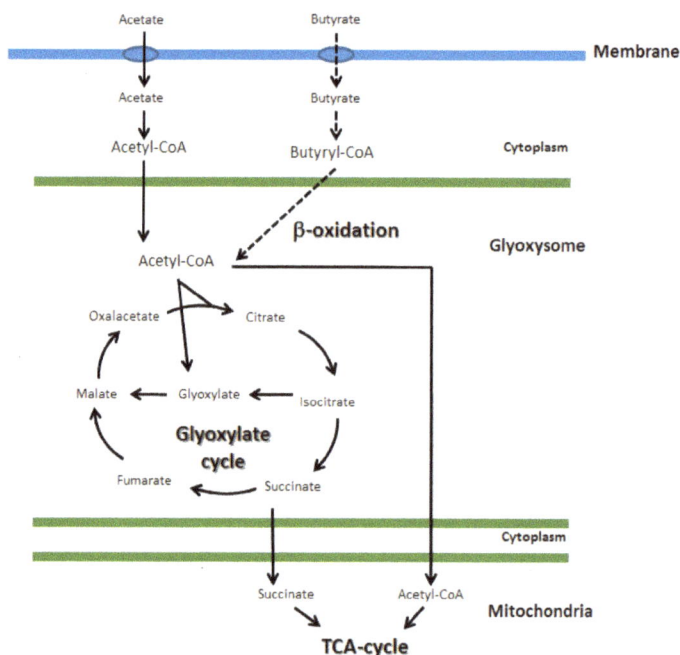

Figure 1. Scheme of metabolic pathways for assimilation of acetate and butyrate.

Butyrate is another major by-product of dark fermentation, but butyrate assimilation by microalgae has not yet been studied in such detail. Butyrate has relatively higher molecular weight and associated complicity and requires more steps for conversion to acetyl-CoA. Similar to transport of acetate, it can be anticipated that butyrate enters via a monocarboxylic/proton transporter across the membrane. In the glyoxysome, butyrate is converted to acetyl-CoA through β-oxidation [43].

The acetyl-CoA is partially used via the glyoxylate cycle for biosynthesis, as well as via the TCA cycle for energy production (Figure 1). For both metabolic cycles, acetyl-CoA is a prerequisite and is mainly provided by β-oxidation [46]. Although acetate can be efficiently converted into lipids, butyrate uptake by microalgae is much slower and can reduce the microalgae growth when both VFA are present. This problem can be solved either by increasing the initial microalgae biomass or by increasing the initial acetate: butyrate ratio [31].

3.2. Diverse VFA as a Carbon Source

Until now, only a limited number of microalgae have been reported in literature to utilize VFA for their growth (Table 2). Various carbon sources have been reported including one only type or mixtures of VFA that include acetate, propionate and butyrate with acetate to be the most commonly used. The process followed for VFA utilization is mainly heterotrophic or mixotrophic, while in one only study, photoheterotrophic fermentation of *C. vulgaris* was carried out [38]. In most cases, VFA were utilized for lipid production with *Crypthecodinium cohnii* to be the most important player in omega-3 production, while in rare cases VFA was utilized for the production of microalga biomass. Although some microalgae are capable of producing carotenoids, according to literature so far VFA utilization has been linked only with lipid or biomass and no carotenoid production. An exemption to this trend is the production of astaxanthin by *Haematococcus pluvialis* that has the potential to utilize acetate in a heterotrophic manner [47]. However, the maximum astaxanthin content of *H. pluvialis* cells was reported to be lower than 3-fold compared to the production carried out by photoautotrophic process.

Table 2. Microalgal strains that utilize VFA as carbon source.

Strain	Carbon Source	Process	Product	Production	Ref.
Chlamydomonas reinhardtii wild-type strain CC-124	Acetate	mixotrophic	Microalgal oil	16.41 ± 1.12% lipid content	[48]
Chlamydomonas reinhardtii wild-type strain CC-124	acetic acid: propionic acid: butyric acid in a ratio 8:1:1	mixotrophic	Microalgal oil	19.02% lipid content	[48]
Chlorella protothecoides 249	Acetate	mixotrophic, initial pH = 6.5	Microalgal oil *	29.45 ± 0.84% lipid content	[49]
Chlorella protothecoides 249	Acetate	heterotrophic; initial pH = 6.3	Microalgal oil	52.38 ± 25.77% lipid content	[49]
Chlorella protothecoides FACHB-3	WAS hydrolysate containing a mixture of VFA	heterotrophic	Microalgal oil	21.5 ± 1.44% lipid content	[50]
Chlorella protothecoides UTEX 25	acetic acid: propionic acid: butyric acid in a ratio 8:1:1	Heterotrophic (nitrogen source = Urea)	Microalgal oil	48.7 ± 2.2 % lipid content (0.317 ± 0.01 g/L)	[51]
Chlorella sorokiniana (CCAP 211/8K)	acetate and butyrate	mixotrophic	Microalga biomass	1.14 g CDW/L	[52]
Chlorella sp. (Arctic) ArM0029B	acetic, propionic and butyric acids in ratio 6:1:3 (by mass)	CO2 supplied-mixotrophic followed by the VFA supply	Microalgal oil	65.7 ± 3.1 mg/g CDW	[53]
Chlorella vulgaris ESP6	mixture of acetate, lactate, butyrate, and HCO-	photoheterotrophic, constant pH = 7.5	Microalga biomass	0.87 g CDW/L	[38]
Chlorella vulgaris UTEX 259	Acetate	mixotrophic	Microalgal oil	36 ± 1% lipid content	[54]
Crypthecodinium cohnii strain ATCC 30772	Acetate	heterotrophic (pH-auxostat culture)	DHA	4.4 g/L	[55]
Ettlia sp. YC001	acetic, propionic and butyric acids in ratio 6:1:3 (by mass)	CO2 supplied-mixotrophic followed by the VFA supply	Microalgal oil	88.5 ± 0.0 mg/g CDW	[53]
Haematococcus pluvialis NIES-144	Acetate	heterotrophic	astaxanthin	22.6 mg/g CDW	[47]
Micractinium inermum F014	acetic, propionic and butyric acids in ratio 6:1:3 (by mass)	CO2 supplied-mixotrophic followed by the VFA supply	Microalgal oil	62.6 ± 0.5 mg/g CDW	[53]
Scenedesmus obliquus (UTEX 78)	Acetate	mixotrophic	Microalga biomass	>1.4 day^{-1} growth rate	[56]
Scenedesmus sp. strain R-16	Acetate	heterotrophic	Microalgal oil	34.4% lipid content	[57]
Scenedesmus sp. strain R-16	Butyrate	heterotrophic	Microalgal oil	24.8% lipid content	[57]

* Total fatty acids. CDW = Cell Dry Weight.

3.2.1. Acetate as a Single Carbon Source

Acetic acid is the most abundant of VFA and the one most easily produced. As already mentioned, some microalgae have exhibited the ability to grow heterotrophically on acetate. *Scenedesmus obliquus* has been found to increase its growth rate under mixotrophic conditions, with the addition of acetate as carbon source, in comparison to the autotrophic one [56]. Also certain *Chlorella* species are known to grow on acetate as sole carbon source and produce various metabolites [54,58]. In those species, mixotrophic conditions usually favor the acetate uptake and biomass increase more than the heterotrophic ones [48]. This can be attributed to the ability of light to boost the biomass production of microalgae at the starting point of the culture, thus producing biomass and therefore accelerating the apparent heterotrophic uptake of VFA.

The most characteristic example of heterotrophic growth on acetate is that of the heterotrophic microalga *Crypthecodinium cohnii*. The alga can grow rapidly on acetate in a pH-auxostat were pH is controlled by addition of acetic acid [55]. This specific microorganism is industrially cultivated for production of DHA.

3.2.2. Mixture of VFA as Carbon Source

Apart from acetate, the sole use of butyrate as carbon source has been proposed for heterotrophic or mixotrophic cultivation of microalgae. However, various experiments have showed that butyrate seems to possess an extremely low ability of being assimilated by microalgae [42]. Furthermore, the addition of butyrate together with acetate in the culture broth has been found to act as an inhibitor and slow down the acetate uptake [52]. Therefore, in order to better understand the behavior of microalgae under heterotrophic conditions, we need to examine their ability to grow, not only on a sole VFA as carbon source, but also on mixtures of different composition [43].

Different microalgal species, when grown on a mixture of VFA, consume acetate preferably to the other acids such us butyric or propionic. Therefore, it is a general rule that a higher acetate proportion in the VFA mixture provides higher lipid accumulation and biomass growth. For example, both *Chlamydomonas reinhardtii* and *Chlorella protothecoides* were best cultivated in a mixture of acetic, propionic and butyric acid with a ratio of 8:1:1 [48,51]. Another usual ratio for microalgae cultivation, in which acetate is the most abundant acid, is 6:1:3 [53]. The main reason for this ratio preference appears to be the tendency of microalgae to exhibit a diauxic growth behaviour when the culture medium contains a mixture of carbon sources. Initially, only the substrate that supports the highest growth rate is utilized by the alga, while the consumption of other substrates remains repressed. Only after the preferred substrate is totally consumed the microalga begins metabolizing the other carbon sources [59]. For this reason, a high acetate concentration can ensure that enough cells will initially be produced in order to consume the other "poorer" substrates afterwards.

3.2.3. Dark Fermentation Effluents as Carbon Source

VFA are the main by-products of dark fermentation process for the production of hydrogen. Microalgae can consume VFA and produce biomass and useful metabolites, posing therefore a very good solution for the reduction of the acid intermediates. However, the dark fermentation effluents contain many substances, apart from VFA that may act as inhibitors to microalgae growth. The consumption of VFA present in those effluents does not follow necessarily the same rules as the consumption of the previously mentioned VFA mixtures.

The composition of a dark fermentation effluent is hard to be estimated, since it is highly dependent on the composition of the waste and the microbial consortium [8]. So far only a small number of experiments have been made in order to examine the growth of certain microalgal species on such effluents. Cho et al. have managed to cultivate *C. vulgaris* on sewage sludge fermentation effluent mixotrophically. It was concluded that the various substances of the effluent did not act as inhibitors to the growth of the alga [60]. In addition, Wen et al. managed to heterotrophically grow *C. prototfecoides* on WAS hydrolysate, that contained high amounts of VFA, supplemented with nutrients [50]. It has also been suggested that microalgae grew well in fed-batch cultivation on raw effluents since the slow addition of the effluent caused the VFA concentration to remain lower than the inhibitory level [10].

3.2.4. Effect of Culture Conditions in VFA Utilization

Apart from the VFA ratio and the provision of light, various other culture conditions should be taken into consideration in order for the microalgae to grow. The pH of the culture broth seems to be an important factor. Although the appropriate pH of each microalga species differs, it has been repeatedly found that controlling the pH of the process results in higher biomass and product yields than when allowing it to vary naturally [38,50,61]. Under these conditions it is more difficult for pH-related inhibition, caused by the uptake of VFA to be developed, such as cytosolic pH acidification [62].

The optimal temperature for microalgae growth can enhance the enzymatic activity of several cell proteins, as well as reduce the energy requirements for thermoregulation. However, sometimes lower temperatures that enhance the production of a certain metabolite, due to stress conditions, are adopted. Higher growth temperatures than the optimal cause problems to the culture, especially in case of a mixotrophic process. The oxidase activity of many enzymes, including the ribulose-1,5-bisphosphate carboxylase/oxygenase (RuBisCO) enzyme, in those temperatures rises. As a result, a higher photorespiration rate occurs and an amount of the energy produced during photosynthesis is being wasted [43].

Another very important factor influencing the lipid and biomass growth is the nitrogen source. It is widely accepted that a high C/N ratio can accumulate the production of various metabolites, such as carotenoids and lipids, from microalgae, as a response to stress conditions. Therefore, to optimize the incompatible mechanisms of microalgae growth and lipids accumulation usually a method of nitrogen limitation is adopted. However, the nitrogen source that is used is also a rather determinant factor. More specifically, ammonium seems to be the preferred nitrogen source by most species, since it needs the less energy for its breakdown and uptake [3]. After ammonium, organic substrates like urea and glycine are usually more suitable nitrogen sources than inorganic ones [51,61].

Finally, sometimes, also the culture medium plays a role at the biomass growth [63]. For example, Liu et al. discovered that *C. zofingiensis* could grow faster mixotrophically on Bold's Basal medium with acetate than on Tris-Acetate-Phosphate (TAP) medium, because it could also utilize the inorganic carbon salts [38].

3.3. Production of Bioactive Compounds from Heterotrophic Microalgae Cultivation

Production of microalgae has gained a lot of interest in the past decades. Due to their biodiversity, as well as the possibility to manipulate biochemical composition by changing culture conditions, microalgae can synthesize various bioactive chemicals with potential industrial applications. In this context, heterotrophic microalgae cultivation is very attractive for obtaining high added-value products from cellular storage compounds, such as lipids and starch, while being more flexible compared to autotrophic cultivation. Especially, heterotrophic algae naturally have oil content higher than phototrophic ones with a better nutritional profile in fatty acids.

3.3.1. Omega-3 Fatty Acids

According to European Food Safety Authority (EFSA-Q-2004-107), there are two main categories of omega-3 fatty acids (n-3 polyunsaturated fatty acids, PUFAs), which differ in function and requirements; the first refers to α-linolenic acid (ALA) produced from vegetable oil, while the other includes long chain n-3 polyunsaturated fatty acids (LC n-3 PUFAs) from marine sources. LC-PUFAs, such as EPA and DHA, have been widely recognized as important bioactive compounds that can be used in the food and nutraceutical industry for the development of functional foods with scientifically proven benefits [64]. There have been reported many studies related to health benefits of DHA and

EPA. ALA is also nutritionally essential and is required for the synthesis of important fatty acids and eicosanoids. Nevertheless, the conversion of ALA to long chain PUFAs depends on several factors, such as the concentration of omega-6 fatty acids. Available evidence suggests that LC n-3 PUFA (EPA and DHA) may reduce the risk of cardiovascular disease, possibly mediated by prevention of cardiac arrhythmias [65]. Positive effects on blood lipid level can be found if the diet contains long chain polyunsaturated omega-3 fatty acids. Especially, EPA and DHA are associated with reduced risk of cardiovascular diseases due to their positive effect on plasma triglyceride levels. Other potential mechanisms of LC n-3 PUFA on cardiovascular protection are related to the capability to reduce blood pressure, their anti-inflammatory and antiarrhythmic effects as well as lowering thrombotic tendency. Furthermore, improved vascular endothelial function and insulin sensitivity as well as, increased plaque stability and paraoxonase levels are linked to polyunsaturated omega-3 fatty acids [66].

PUFAs have vital structural and functional roles in higher organisms, including human, and are related to prevention of cardiovascular and inflammatory diseases, cancer and diabetes. Fish oil is considered to be the conventional source of PUFAs; however, the process encounters several limitations concerning the reducing fish populations or the presence of contaminants, like dioxins, polychlorinated biphenyls and heavy metals. On the basis of increasing global fish meal and fish oil costs, it is predicted that dietary fish meal and fish oil inclusion levels within compound aqua feeds will decrease in the long term, resulting in lower levels of healthy omega-3 fatty acids in aquaculture produced fish [67]. Therefore, new sources and production systems, taking into consideration sustainability and economic feasibility for long chain polyunsaturated fatty acids, must be developed for direct human consumption as well as for aquaculture. The main source of omega-3 is still fish oil. Since only plants are able to synthesize essential fatty acids, microalgae could supply the whole food chain with these very important components. In addition to fatty acids, algae contain a massive number of very valuable substances and chemical compounds such as lipids and polysaccharides, sterols, phycobiliproteins or antioxidants such as tocopherol (vitamin E), ascorbic acid (vitamin C), carotenoids and the red pigment astaxanthin.

Microalgae are able to accumulate high amounts of EPA and DHA when growing in heterotrophic cultures, supported by a carbon source [68]. The lipid content in heterotrophically cultivation of *C. protothecoides* has been reported to reach 55 wt. %, which is 4-times greater than the autotrophically grown cultures under similar conditions [69]. The production of PUFAs is commonly observed during the stationary phase, when the cells have most of their biosynthetic capacities redirected to the production of lipids [70]. At present, industrial production of heterotrophic microalgae is hampered by the high economic and environmental costs of glucose, commonly used as main carbon source. Using different waste streams within the frame of a VFA platform as potential substrates, is a promising strategy for developing a sustainable bio-economy.

Although VFA and especially acetate can successfully support microalgal growth, they do not serve to induce lipid production in the cells. In order to do so, an enviromental stress is required. The usual technique that is adopted is cultivation under nitrogen starvation, that is known to enhance the production of lipids in form of triacylglycerides [71]. This, however, means that the accumulation of lipids antagonizes the biomass production. Only very recently, the discovery of some small molecular activators, that enhance the lipid accumulation, without hampering the biomass growth has been made [72].

3.3.2. Production of Other High Value Added Compounds

Several algal strains have been explored under heterotrophic cultivation to develop their capacity in producing not only DHA and EPA, but also interesting and high demanded value-added compounds, such as carotenoids, phycobiliproteins, polysaccharides and others. *Thraustochytrids* strains (*Aurantiochytrium* sp., *Schizochytrium* sp., *Thraustochytrium* sp., *Ulkenia* sp.) have been evaluated for production of biodiesel, long-chain omega-3 oils and exopolysaccharides (EPS), concluding that *Aurantiochytrium* sp. is the best candidate for production of biofuels, PUFAs and EPS [66]. Other algal strains like *C. protothecoides*, *Galdieria sulphuraria*, *Nitzchia laevis*, *C. cohnii* and *Neochloris oleabundans* [37], have been studied due to their content in hydrogen, lipids and carotenoids, presenting a great potential to be used at large scale. *C. zofingiensis* has been examined as a heterotrophic alternative to the microalga *H. pluvialis* for production of the carotenoid astaxanthin [73]. It should be mentioned that only a few bioactive compounds apart from lipids, such as astaxanthin and β-carotene, have been produced at industrial scale, due to the low production yields in microalgae cells and the difficulties in isolating/purifying them by economically feasible processes [74,75]. In order to further utilize microalgae advantages as means of bioconversion for more useful products, extensive research is being conducted on optimization of cultivation conditions and genetic engineering providing strains with high productivity yields of bioactive compounds.

3.3.3. Downstream Processing

Downstream processing involves several technologies that cover biomass harvesting, cell disruption and extraction and the final separation and purification step. In these processes are involved key factors that should be taken into account when an industrial scale up is being performed, such as low energy costs, scalability and specially a critical factor, all useful compounds like carbohydrates, pigments, ω-3 fatty acids, proteins might maintain their functionality after downstream processing. Despite the efforts made during recent years, the downstream processing step is still one of the main responsible of the high cost of microalgae production at industrial scale [70]. In this context, heterotrophic cultivation system shows an interesting advantage due to its capacity to growth under all type of luminous conditions. Some studies have been demonstrated that the independence of light to growth is translated to higher biomass yield and consequently downstream processing steps are reduced. Under heterotrophic conditions, it is shown that ω-3 fatty acid concentration can be two or three orders of magnitude greater that those under autotrophic conditions [76].

From an economical point of view, harvesting process is one of the major bottlenecks for commercialization of products from microalgae. Many efforts are being done to improve this step and it is one of the most challenging areas in the algal biofuel research. Current harvesting technologies applied include chemical, electrical, mechanical and biological based methods. In particular, cell flocculation technology to recover biomass from microalgae is becoming of great interest due to its capacity for treating large amounts of biomass with low energy cost [76,77]. In this context, bio-flocculation and auto-flocculation as novel flocculation based methods are highlighted among others, as it could be shown recently. Nevertheless, there are still some technical barriers to decrease costs and complexity of downstream processing in microalgae production that might be solved aiming future industrial applications [78,79].

On the other hand, as previously mentioned, extraction and purification steps might maintain the functionality of compounds of interest. It is critical in the case of carotenoids and lipids due to their facility for oxidizing in presence of light and oxygen. Using heterotrophic microalgae cultivation for biodiesel production, lipid content is the key factor and therefore special attention is being taken to reduce costs in this step. Traditionally organic solvent extraction was the most common method used with a lipid yield around 15–20% depending on the solvent. Additional technologies were applied in combination with organic solvent extraction to improve lipid yield extraction such as microwave, sonication and homogenization. However, these methods result in higher energy consumption without the possibility of microalgae re-cultivation after solvent extraction due to the toxicity of solvents used. The potential of the fatty acids being used as a food additive usually prohibits the use of toxic solvents. Therefore, less-toxic, but unfortunately less effective, solvents, such as isopropanol, butanol, methyl tert-butyl ether (MTBE), acetic acid esters, hexane, 2-ethoxyethanol (2-EE) or ethanol have been also tested for microalgal lipid extraction [80]. Recently novel technologies are being studied, such as nanotechnology application, not only in the extraction but also in final transesterification step. A comprehensive description of the technology and different nanomaterials applied is reviewed by Zhang et al. (2013) [81], proving that nanomaterials are capable of efficiently extracting compounds from the cells, without any impact on microalgae growth. The studies of nanotechnology application in biodiesel production are still in lab-scale but preliminary results render this field promising and further research should be made in this direction.

4. Discussion and Future Perspectives

It appears that microalgae are a very promising type of biorefinery. Their advantage lies, not only on their ability to produce various high-added value metabolites, but also on the versatility that characterizes their growth modes. Microalgae can grow heterotrophically, as well as autotrophically, and even utilize carbon sources such as VFA. VFA are environmentally harmful by-products of dark fermentation process for the production of hydrogen from organic waste. Their utilization by heterotrophic microalgae for industrial production of lipids and carotenoids solves the economic problem that arises from the high cost of glucose as a carbon source. As it is proven, some microalgae genera such as *Chlamydomonas*, *Chlorella*, *Crypthecodinium* and *Scenedesmus* can grow well on VFA mixtures, even on dark fermentation effluents, with no significant inhibitory effects. In order to broaden our knowledge on the subject and establish an industrial method of utilizing VFA for microalgal metabolites production, attempts on further research are highly encouraged.

Acknowledgments: Angelina Chalima, Laura Oliver, Laura Fernández de Castro, Thomas Dietrich and Evangelos Topakas would like to thank VOLATILE, a project funded by the European Union's Horizon 2020 research and innovation program, under the grant agreement no. 720777. Anthi Karnaouri wishes to acknowledge financial support by the Greek State Scholarships (Postdoc-Research Scholarships IKY). Angelina Chalima would like to thank the Hellenic Foundation of Research and Innovation (ELIDEK) for financial support (ELIDEK Scholarships for Ph.D. Students).

Author Contributions: Evangelos Topakas and Thomas Dietrich conceived the content and structure of manuscript; Angelina Chalima, Laura Oliver, Laura Fernández de Castro and Anthi Karnaouri analyzed the literature data and wrote the paper.

Conflicts of Interest: The authors declare no conflict of interest.

References

1. Lee, W.S.; Chua, A.S.M.; Yeoh, H.K.; Ngoh, G.C. A review of the production and applications of waste-derived volatile fatty acids. *Chem. Eng. J.* **2014**, *235*, 83–99. [CrossRef]
2. Chen, Y.; Jiang, X.; Xiao, K.; Shen, N.; Zeng, R.J.; Zhou, Y. Enhanced volatile fatty acids (VFAs) production in a thermophilic fermenter with stepwise pH increase—Investigation on dissolved organic matter transformation and microbial community shift. *Water Res.* **2017**, *112*, 261–268. [CrossRef] [PubMed]
3. Perez-Garcia, O.; Escalante, F.M.E.; de-Bashan, L.E.; Bashan, Y. Heterotrophic cultures of microalgae: Metabolism and potential products. *Water Res.* **2011**, *45*, 11–36. [CrossRef] [PubMed]
4. Venkata Mohan, S.; Prathima Devi, M. Fatty acid rich effluent from acidogenic biohydrogen reactor as substrate for lipid accumulation in heterotrophic microalgae with simultaneous treatment. *Bioresour. Technol.* **2012**, *123*, 627–635. [CrossRef] [PubMed]
5. Sinha, B. Omega-3 Market by Type (ALA, EPA, and DHA), Source (Fish Oil & Krill Oil, Algal Oil, Walnut, Pumpkin Seeds, Soybean Oil, Canola Oil, Bean Curd, and Others), Application (Dietary Supplement, Pharmaceutical, Infant Formula, Food & Beverage, Pet Food, and Fish Feed)-Global Opportunity Analysis and Industry Forecast, 2014–2022. Allied Market Research: Portland, OR, USA, 2016. Available online: https://www.alliedmarketresearch.com/omega-3-market (accessed on 29 August 2016).
6. Sprague, M.; Dick, J.R.; Tocher, D.R. Impact of sustainable feeds on omega-3 long-chain fatty acid levels in farmed Atlantic salmon, 2006–2015. *Sci. Rep.* **2016**, *6*, 21892. [CrossRef] [PubMed]
7. Huang, Y.L.; Wu, Z.; Zhang, L.; Cheung, C.M.; Yang, S.T. Production of carboxylic acids from hydrolyzed corn meal by immobilized cell fermentation in a fibrous-bed bioreactor. *Bioresour. Technol.* **2002**, *82*, 51–59. [CrossRef]
8. Arudchelvam, Y.; Perinpanayagam, M.; Nirmalakhandan, N. Predicting VFA formation by dark fermentation of particulate substrates. *Bioresour. Technol.* **2010**, *101*, 7492–7499. [CrossRef] [PubMed]

9. Singhania, R.R.; Patel, A.K.; Christophe, G.; Fontanille, P.; Larroche, C. Biological upgrading of volatile fatty acids, key intermediates for the valorization of biowaste through dark anaerobic fermentation. *Bioresour. Technol.* **2013**, *145*, 166–174. [CrossRef] [PubMed]
10. Hu, B.; Zhou, W.; Min, M.; Du, Z.; Chen, P.; Ma, X.; Liu, Y.; Lei, H.; Shi, J.; Ruan, R. Development of an effective acidogenically digested swine manure-based algal system for improved wastewater treatment and biofuel and feed production. *Appl. Energy* **2013**, *107*, 255–263. [CrossRef]
11. Yan, Y.; Feng, L.; Zhang, C.; Wisniewski, C.; Zhou, Q. Ultrasonic enhancement of waste activated sludge hydrolysis and volatile fatty acids accumulation at pH 10.0. *Water Res.* **2010**, *44*, 3329–3336. [CrossRef] [PubMed]
12. Pham, T.N.; Nam, W.J.; Jeon, Y.J.; Yoon, H.H. Volatile fatty acids production from marine macroalgae by anaerobic fermentation. *Bioresour. Technol.* **2012**, *124*, 500–503. [CrossRef] [PubMed]
13. Fu, Z.; Holtzapple, M.T. Consolidated bioprocessing of sugarcane bagasse and chicken manure to ammonium carboxylates by a mixed culture of marine microorganisms. *Bioresour. Technol.* **2010**, *101*, 2825–2836. [CrossRef] [PubMed]
14. Han, W.; Chen, H.; Jiao, A.; Wang, Z.; Li, Y.; Ren, N. Biological fermentative hydrogen and ethanol production using continuous stirred tank reactor. *Int. J. Hydrogen Energy* **2012**, *37*, 843–847. [CrossRef]
15. Ding, H.B.; Tan, G.Y.A.; Wang, J.Y. Caproate formation in mixed-culture fermentative hydrogen production. *Bioresour. Technol.* **2010**, *101*, 9550–9559. [CrossRef] [PubMed]
16. Teghammar, A.; Yngvesson, J.; Lundin, M.; Taherzadeh, M.J.; Horváth, I.S. Pretreatment of paper tube residuals for improved biogas production. *Bioresour. Technol.* **2010**, *101*, 1206–1212. [CrossRef] [PubMed]
17. Wikandari, R.; Gudipudi, S.; Pandiyan, I.; Millati, R.; Taherzadeh, M.J. Inhibitory effects of fruit flavors on methane production during anaerobic digestion. *Bioresour. Technol.* **2013**, *145*, 188–192. [CrossRef] [PubMed]
18. Zhou, A.; Du, J.; Varrone, C.; Wang, Y.; Wang, A.; Liu, W. VFAs bioproduction from waste activated sludge by coupling pretreatments with *Agaricus bisporus* substrates conditioning. *Process Biochem.* **2014**, *49*, 283–289. [CrossRef]
19. Xu, G.; Chen, S.; Shi, J.; Wang, S.; Zhu, G. Combination treatment of ultrasound and ozone for improving solubilization and anaerobic biodegradability of waste activated sludge. *J. Hazard. Mater.* **2010**, *180*, 340–346. [CrossRef] [PubMed]
20. Zhou, A.; Yang, C.; Guo, Z.; Hou, Y.; Liu, W.; Wang, A. Volatile fatty acids accumulation and rhamnolipid generation in situ from waste activated sludge fermentation stimulated by external rhamnolipid addition. *Biochem. Eng. J.* **2013**, *77*, 240–245. [CrossRef]
21. Kim, H.J.; Choi, Y.G.; Kim, G.D.; Kim, S.H.; Chung, T.H. Effect of enzymatic pretreatment on solubilization and volatile fatty acid production in fermentation of food waste. *Water Sci. Technol.* **2005**, *52*, 51–59. [CrossRef] [PubMed]
22. Kim, H.J.; Kim, S.H.; Choi, Y.G.; Kim, G.D.; Chung, T.H. Effect of enzymatic pretreatment on acid fermentation of food waste. *J. Chem. Technol. Biotechnol.* **2006**, *81*, 947–980. [CrossRef]
23. Bayr, S.; Kaparaju, P.; Rintala, J. Screening pretreatment methods to enhance thermophilic anaerobic digestion of pulp and paper mill wastewater treatment secondary sludge. *Chem. Eng. J.* **2013**, *223*, 479–486. [CrossRef]
24. Yang, Q.; Luo, K.; Li, M.X.; Bo Wang, D.; Zheng, W.; Zeng, M.G.; Liu, J.J. Enhanced efficiency of biological excess sludge hydrolysis under anaerobic digestion by additional enzymes. *Bioresour. Technol.* **2010**, *101*, 2924–2930. [CrossRef] [PubMed]
25. Golub, K.W.; Smith, A.D.; Hollister, E.B.; Gentry, T.J.; Holtzapple, M.T. Investigation of intermittent air exposure on four-stage and one-stage anaerobic semi-continuous mixed-acid fermentations. *Bioresour. Technol.* **2011**, *102*, 5066–5075. [CrossRef] [PubMed]
26. Borzacconi, L.; Lopez, I.; Anido, C. Hydrolysis constant and VFA inhibition in acidogenic phase of MSW anaerobic degradation. *Water Sci. Technol.* **1997**, *36*, 479–484. [CrossRef]
27. Zhou, A.; Guo, Z.; Yang, C.; Kong, F.; Liu, W.; Wang, A. Volatile fatty acids productivity by anaerobic co-digesting waste activated sludge and corn straw: Effect of feedstock proportion. *J. Biotechnol.* **2013**, *168*, 234–239. [CrossRef] [PubMed]
28. Feng, L.; Chen, Y.; Zheng, X. Enhancement of waste activated sludge protein conversion and volatile fatty acids accumulation during waste activated sludge anaerobic fermentation by carbohydrate substrate addition: The effect of pH. *Environ. Sci. Technol.* **2009**, *43*, 4373–4380. [CrossRef] [PubMed]

29. Yu, H.Q.; Fang, H.H.P. Acidification of mid- and high-strength dairy wastewaters. *Water Res.* **2001**, *35*, 3697–3705. [CrossRef]

30. Pessiot, J.; Nouaille, R.; Jobard, M.; Singhania, R.R.; Bournilhas, A.; Christophe, G.; Fontanille, P.; Peyret, P.; Fonty, G.; Larroche, C. Fed-batch anaerobic valorization of slaughterhouse by-products with mesophilic microbial consortia without methane production. *Appl. Biochem. Biotechnol.* **2012**, *167*, 1728–1743. [CrossRef] [PubMed]

31. Turon, V.; Trably, E.; Fayet, A.; Fouilland, E.; Steyer, H.Q. Raw dark fermentation effluent to support heterotrophic microalgae growth: Microalgae successfully outcompete bacteria for acetate. *Algal Res.* **2015**, *12*, 119–125. [CrossRef]

32. Yu, H.Q.; Fang, H.H.P. Acidogenesis of gelatin-rich wastewater in an upflow anaerobic reactor: Influence of pH and temperature. *Water Res.* **2003**, *37*, 55–66. [CrossRef]

33. Ji, Z.; Chen, G.; Chen, Y. Effects of waste activated sludge and surfactant addition on primary sludge hydrolysis and short-chain fatty acids accumulation. *Bioresour. Technol.* **2010**, *101*, 3457–3462. [CrossRef] [PubMed]

34. Chang, H.N.; Kim, N.J.; Kang, J.; Jeong, C.M. Biomass-derived volatile fatty acid platform for fuels and chemicals. *Biotechnol. Bioprocess Eng.* **2010**, *15*, 1–10. [CrossRef]

35. Bengtsson, S.; Hallquist, J.; Werker, A.; Welander, T. Acidogenic fermentation of industrial wastewaters: Effects of chemostat retention time and pH on volatile fatty acids production. *Biochem. Eng. J.* **2008**, *40*, 492–499. [CrossRef]

36. Huang, L.; Chen, B.; Pistolozzi, M.; Wu, Z.; Wang, J. Inoculation and alkali coeffect in volatile fatty acids production and microbial community shift in the anaerobic fermentation of waste activated sludge. *Bioresour. Technol.* **2014**, *153*, 87–94. [CrossRef] [PubMed]

37. Jie, W.; Peng, Y.; Ren, N.; Li, B. Volatile fatty acids (VFAs) accumulation and microbial community structure of excess sludge (ES) at different pHs. *Bioresour. Technol.* **2014**, *152*, 124–129. [CrossRef] [PubMed]

38. Liu, C.H.; Chang, C.Y.; Liao, Q.; Zhu, X.; Chang, J.S. Photoheterotrophic growth of *Chlorella vulgaris* ESP6 on organic acids from dark hydrogen fermentation effluents. *Bioresour. Technol.* **2013**, *145*, 331–336. [CrossRef] [PubMed]

39. Morales-Sánchez, D.; Martinez-Rodriguez, O.A.; Kyndt, J.; Martinez, A. Heterotrophic growth of microalgae: Metabolic aspects. *World J. Microbiol. Biotechnol.* **2015**, *31*, 1–9. [CrossRef] [PubMed]

40. Barclay, W.R.; Meager, K.M.; Abril, J.R. Heterotrophic production of long chain omega - 3 fatty acids utilizing algae and algae - like microorganisms. *J. Appl. Phycol.* **1994**, *6*, 123–129. [CrossRef]

41. Pérez-Garcia, O.; Bashan, Y. Microalgal heterothophic and mixotrophic culturing for bio-refining: From metabolic routes to techno-economics. In *Algal Biorefineries*; Prokop, A., Bajpai, R.K., Zappi, M.E., Eds.; Springer International Publishing: Cham, Switzerland, 2015; pp. 61–131. [CrossRef]

42. Turon, V.; Baroukh, C.; Trably, E.; Latrille, E.; Fouilland, E.; Steyer, J.P. Use of fermentative metabolites for heterotrophic microalgae growth: Yields and kinetics. *Bioresour. Technol.* **2015**, *175*, 342–349. [CrossRef] [PubMed]

43. Turon, V.; Trably, E.; Fouilland, E.; Steyer, J.P. Potentialities of dark fermentation effluents as substrates for microalgae growth: A review. *Process Biochem.* **2016**, *51*, 1843–1854. [CrossRef]

44. Miller, R.; Wu, G.; Deshpande, R.R.; Vieler, A.; Gärtner, K.; Li, X.; Moellering, E.R.; Zäuner, S.; Cornish, A.J.; Liu, B.; et al. Changes in transcript abundance in *Chlamydomonas reinhardtii* following nitrogen deprivation predict diversion of metabolism. *Plant Physiol.* **2010**, *154*, 1737–1752. [CrossRef] [PubMed]

45. Wase, N.; Black, P.N.; Stanley, B.A.; DiRusso, C.C. Integrated quantitative analysis of nitrogen stress response in *Chlamydomonas reinhardtii* using metabolite and protein profiling. *J. Proteome Res.* **2014**, *13*, 1373–1396. [CrossRef] [PubMed]

46. Kunze, M.; Pracharoenwatana, I.; Smith, S.M.; Hartig, A. A central role for the peroxisomal membrane in glyoxylate cycle function. *Biochim. Biophys. Acta* **2006**, *1763*, 1441–1452. [CrossRef] [PubMed]

47. Kang, C.D.; Lee, J.S.; Park, T.H.; Sim, S.J. Comparison of heterotrophic and phtoautotrophic induction on astaxanthin production by *Haematococcus pluvialis*. *Appl. Microbiol. Biotechnol.* **2005**, *68*, 237–241. [CrossRef] [PubMed]

48. Moon, M.; Kim, C.W.; Park, W.K.; Yoo, G.; Choi, Y.E.; Yang, J.W. Mixotrophic growth with acetate or volatile fatty acids maximizes growth and lipid production in *Chlamydomonas reinhardtii*. *Algal Res.* **2013**, *2*, 352–357. [CrossRef]

49. Heredia-Arroyo, T.; Wei, W.; Hu, B. Oil accumulation via heterotrophic/mixotrophic *Chlorella protothecoides*. *Appl. Biochem. Biotechnol.* **2010**, *162*, 1978–1995. [CrossRef] [PubMed]

50. Wen, Q.; Chen, Z.; Li, P.; Duan, R.; Ren, N. Lipid production for biofuels from hydrolyzate of waste activated sludge by heterotrophic *Chlorella protothecoides*. *Bioresour. Technol.* **2013**, *143*, 695–698. [CrossRef] [PubMed]

51. Fei, Q.; Fu, R.; Shang, L.; Brigham, C.J.; Chang, H.N. Lipid production by microalgae *Chlorella protothecoides* with volatile fatty acids (VFAs) as carbon sources in heterotrophic cultivation and its economic assessment. *Bioprocess Biosyst. Eng.* **2015**, *38*, 691–700. [CrossRef] [PubMed]

52. Turon, V.; Trably, E.; Fouilland, E.; Steyer, J.P. Growth of *Chlorella sorokiniana* on a mixture of volatile fatty acids: The effects of light and temperature. *Bioresour. Technol.* **2015**, *198*, 852–860. [CrossRef] [PubMed]

53. Ryu, B.G.; Kim, W.; Heo, S.W.; Kim, D.; Choi, G.G.; Yang, J.W. Advanced treatment of residual nitrogen from biologically treated coke effluent by a microalga-mediated process using volatile fatty acids (VFAs) under stepwise mixotrophic conditions. *Bioresour. Technol.* **2015**, *191*, 488–495. [CrossRef] [PubMed]

54. Syrett, P.J.; Bocks, S.M.; Merrett, M.J. The Assimilation of Acetate by *Chlorella vulgaris*. *J. Exp. Bot.* **1969**, *15*, 35–47. [CrossRef]

55. Ratledge, C.; Kanagachandran, K.; Anderson, A.J.; Grantham, D.J.; Stephenson, J.C. Production of docosahexaenoic acid by *Crypthecodinium cohnii* grown in a pH-auxostat culture with acetic acid as principal carbon source. *Lipids* **2001**, *36*, 1241–1246. [CrossRef]

56. Combres, C.; Laliberte, G.; Reyssac, J.S.; de la Noue, J. Effect of acetate on growth and ammonium uptake in the microalga *Scenedesmus obliquus*. *Physiol. Plant* **1994**, *91*, 729–734. [CrossRef]

57. Ren, H.Y.; Liu, B.-F.; Ma, C.; Zhao, L.; Ren, N.Q. A new lipid-rich microalga *Scenedesmus* sp. strain R-16 isolated using Nile red staining: Effects of carbon and nitrogen sources and initial pH on the biomass and lipid production. *Biotechnol. Biofuels* **2013**, *6*, 143. [CrossRef] [PubMed]

58. Matsuka, M.; Miyachi, S.; Hase, E. Acetate metabolism in the process of 'acetate-bleaching' of *Chlorella protothecoides*. *Plant Cell Physiol.* **1969**, *538*, 527–538. [CrossRef]

59. Egli, T.; Lendenmann, U.; Snozzi, M. Kinetics of microbial growth with mixtures of carbon sources. *Antonie Van Leeuwenhoek* **1993**, *63*, 289–298. [CrossRef]

60. Cho, H.U.; Kim, Y.M.; Choi, Y.N.; Xu, X.; Shin, D.Y.; Park, J.M. Effects of pH control and concentration on microbial oil production from *Chlorella vulgaris* cultivated in the effluent of a low-cost organic waste fermentation system producing volatile fatty acids. *Bioresour. Technol.* **2015**, *184*, 245–250. [CrossRef] [PubMed]

61. Chen, G.Q.; Chen, F. Growing phototrophic cells without light. *Biotechnol. Lett.* **2006**, *28*, 607–616. [CrossRef] [PubMed]

62. Lin, X.; Xiong, L.; Qi, G.; Shi, S.; Huang, C.; Chen, X.; Chen, X. Using butanol fermentation wastewater for biobutanol production after removal of inhibitory compounds by micro/mesoporous hyper-cross-linked polymeric adsorbent. *ACS Sustain. Chem. Eng.* **2015**, *3*, 702–709. [CrossRef]

63. Caporgno, M.P.; Taleb, A.; Olkiewicz, M.; Font, J.; Pruvost, J.; Legrand, J.; Bengoa, C. Microalgae cultivation in urban wastewater: Nutrient removal and biomass production for biodiesel and methane. *Algal Res.* **2015**, *10*, 232–239. [CrossRef]

64. Khan, R.S.; Grigor, J.; Winger, R.; Win, A. Functional food product development–Opportunities and challenges for food manufacturers. *Trends Food Sci. Technol.* **2013**, *30*, 27–37. [CrossRef]

65. Alabdukarim, B.; Bakeet, Z.A.N.; Arzoo, S. Role of some functional lipids in preventing diseases and promoting health. *J. K. Saud Univ. Sci.* **2012**, *24*, 319–329. [CrossRef]

66. Hooper, L.; Thompson, R.L.; Harrison, R.A.; Summerbell, C.D.; Ness, A.R.; Moore, H.J.; Worthington, H.V.; Durrington, P.N.; Higgins, J.P.T.; Nigel, E.C.; et al. Risks and benefits of omega 3 fats for mortality, cardiovascular disease, and cancer: Systematic review. *Br. Med. J.* **2006**, *332*, 752–755. [CrossRef] [PubMed]

67. Tacon, A.G.J.; Metian, M. Global overview on the use of fish meal and fish oil in industrially compounded aquafeeds: Trends and future prospects. *Aquaculture* **2008**, *285*, 146–158. [CrossRef]

68. Stonik, V.A.; Fedorov, S.N. Marine low molecular weight natural products as potential cancer preventive compounds. *Mar. Drugs* **2014**, *12*, 636–671. [CrossRef] [PubMed]

69. Xu, H.; Miao, X.; Wu, Q. High quality biodiesel production from a microalga *Chlorella protothecoides* by heterotrophic growth in fermenters. *J. Biotechnol.* **2006**, *126*, 499–507. [CrossRef] [PubMed]

70. Hu, Q.; Sommerfeld, M.; Jarvis, E.; Ghirardi, M.; Posewitz, M.; Seibert, M.; Darzins, A. Microalgal triacylglycerols as feedstocks for biofuel production: Perspectives and advances. *Plant J.* **2008**, *54*, 621–639. [CrossRef] [PubMed]
71. Allen, J.W.; DiRusso, C.C.; Black, P.N. Triacylglycerol synthesis during nitrogen stress involves the prokaryotic lipid synthesis pathway and acyl chain remodeling in the microalgae *Coccomyxa subellipsoidea*. *Algal Res.* **2015**, *10*, 110–120. [CrossRef]
72. Wase, N.; Tu, P.; Allen, J.W.; Black, P.N.; DiRusso, C.C. Identification and metabolite profiling of chemical activators of lipid accumulation in green algae. *Plant Physiol.* **2017**. [CrossRef] [PubMed]
73. Liu, J.; Sun, Z.; Gerken, H.; Liu, Z.; Jiang, Y.; Chen, F. *Chlorella zofingiensis* as an alternative microalgal producer of astaxanthin: Biology and industrial potential. *Mar Drugs.* **2014**, *12*, 3487–3515. [CrossRef] [PubMed]
74. Clarens, A.; Resurreccion, E.; White, M.; Colosi, L. Environmental life cycle comparison of algae to other bioenergy feedstocks. *Environ. Sci. Technol.* **2010**, *44*, 1813–1819. [CrossRef] [PubMed]
75. Norsker, N.; Barbosa, M.; Vermue, M.; Wijffels, R. Microalgal production-a close look at the economics. *Biotechnol. Adv.* **2011**, *29*, 24–27. [CrossRef] [PubMed]
76. Vandamme, D.; Foubert, I.; Muylaert, K. Flocculation as a low-cost method for harvesting microalgae for bulk biomass production. *Trends Biotechnol.* **2013**, *31*, 233–239. [CrossRef] [PubMed]
77. Salim, S.; Kosterink, N.R.; Wacka, N.T.; Vermue, M.H.; Wijffels, R.H. Mechanism behind autoflocculation of unicellular green microalgae *Ettliatexensis*. *J. Biotechnol.* **2014**, *174*, 34–38. [CrossRef] [PubMed]
78. Wan, C.; Alam, M.A.; Zhao, X.Q.; Zhang, X.Y.; Guo, S.L.; Ho, S.H.; Chang, J.S.; Bai, F.W. Current progress and future prospect of microalgal biomass harvest using various flocculation technologies. *Bioresour. Technol.* **2015**, *184*, 251–257. [CrossRef] [PubMed]
79. Ummalyma, S.B.; Gnansounou, E.; Sukumaran, R.K.; Sindhy, R.; Pandey, A.; Sahoo, D. Bioflocculation: An alternative strategy for harvesting of microalgae—An overview. *Bioresour. Technol.* **2017**, *242*, 227–235. [CrossRef] [PubMed]
80. Ranjith Kumar, R.; Hanumantha Rao, P.; Arumugam, M. Lipid extraction methods from microalgae: A comprehensive review. *Front. Energy Res.* **2015**, *2*, 1–9. [CrossRef]
81. Zhang, X.L.; Yan, S.; Tyagi, R.D.; Surampalli, R.Y. Biodiesel production from heterotrophic microalgae through transesterification and nanotechnology application in the production. *Renew. Sustain. Energy Rev.* **2013**, *26*, 216–223. [CrossRef]

fermentation

MDPI

Article

Techno-Economic and Life Cycle Assessment of Wastewater Management from Potato Starch Production: Present Status and Alternative Biotreatments

Pedro F. Souza Filho * , Pedro Brancoli , Kim Bolton, Akram Zamani
and Mohammad J. Taherzadeh

Swedish Centre for Resource Recovery, University of Borås, 501 90 Borås, Sweden; pedro.brancoli@hb.se (P.B.);
kim.bolton@hb.se (K.B.); akram.zamani@hb.se (A.Z.); mohammad.taherzadeh@hb.se (M.J.T.)
* Correspondence: pedro.ferreira_de_souza_filho@hb.se; Tel.: +46-70-006-6572; Fax: +46-33-435-4003

Received: 13 September 2017; Accepted: 16 October 2017; Published: 23 October 2017

Abstract: Potato liquor, a byproduct of potato starch production, is steam-treated to produce protein isolate. The heat treated potato liquor (HTPL), containing significant amounts of organic compounds, still needs to be further treated before it is discarded. Presently, the most common strategy for HTPL management is concentrating it via evaporation before using it as a fertilizer. In this study, this scenario was compared with two biotreatments: (1) fermentation using filamentous fungus *R. oryzae* to produce a protein-rich biomass, and (2) anaerobic digestion of the HTPL to produce biogas. Technical, economic and environmental analyses were performed via computational simulation to determine potential benefits of the proposed scenarios to a plant discarding 19.64 ton/h of HTPL. Fungal cultivation was found to be the preferred scenario with respect to the economic aspects. This scenario needed only 46% of the investment needed for the evaporation scenario. In terms of the environmental impacts, fungal cultivation yielded the lowest impacts in the acidification, terrestrial eutrophication, freshwater eutrophication, marine eutrophication and freshwater ecotoxicity impact categories. The lowest impact in the climate change category was obtained when using the HTPL for anaerobic digestion.

Keywords: potato liquor; techno-economic analysis; life cycle assessment; filamentous fungus; anaerobic digestion

---

1. Introduction

Potato is one of the most important food crops in the world, and it accounts for 13.3% of the starch produced in the European Union (EU). The processing of potato to produce starch results in two major byproducts: potato pulp (PP) and potato liquor (PL). PP contains the insoluble polysaccharides cellulose, hemicellulose, pectin and residual starch. Proteins, minerals and trace elements in high concentrations, are the major ingredients of PL [1]. Each metric ton of processed potato yields approximately 200 kg of starch and generates ca 700 kg PL [2] containing 30–41% protein per total solid (TS) [3]. The proteins present in the PL are of good quality, similar to those of whole eggs [3]. Different methods to recover these proteins have been reported. They include thermal coagulation, acid precipitation, salting out, isoelectric precipitation, complexing with carboxymethylcellulose or bentonite, ultrafiltration, expanded bed adsorption and dry separation [3–6]. However, the only method that is presently being used for industrial recovery of protein from PL is heat coagulation [3–7]. In this method, steam is injected into the liquor at a pH of 5.5 to increase the temperature to 99 °C. This coagulates the proteins, which are then precipitated and collected [8]. However, the proteins obtained by this method lose their functional properties and are mostly used as an additive to cattle feed. Moreover, they have a salty, bitter taste, which prevents them from being used as food additives [3,7,9]. After the proteins have been removed from PL, the residual liquid—called HTPL—is further evaporated at 140 °C to produce potato protein liquor (PPL), containing 40% (w/w) TS [1].

The most common method to manage PPL is to use it as fertilizer. However, most potato processing occurs during the winter. The reduced biological activity in the soil during this period prevents the uptake of the liquor, which forces the potato starch facilities to store large volumes of PPL for several months [9]. Additionally, the use of PPL as fertilizer causes contamination of groundwater and emits a bad odor that can disturb citizens living in neighboring areas [10]. However, as mentioned above, despite the high content of residual proteins in the PPL, they are of poor quality, hence preventing them from being used as a supplement for cattle feed [1].

A few alternative management strategies for PPL have been presented in the literature. The biodegradable nature of the byproduct has stimulated its use in bioprocesses to produce yeast biomass [11], acetate and ethanol [12], enzymes [10] and fungal protein [13]. Additionally, the filamentous zygomycete fungus *Rhizopus oryzae* has been used to reduce the chemical oxygen demand (COD) of the PPL and produce a protein-rich biomass with a potential application as fish feed [1].

One of the reasons for the growing interest in producing supplements for fish feed is that aquaculture has been growing at about 8% per year since the late 1970s, which is higher than the rate of human population growth. It has been suggested that the reason for this large growth is the widespread knowledge of the importance of fish for a healthy lifestyle, mainly because fish contains ω-3 polyunsaturated fatty acids. Fish feed is a major cost in intensive fish farming [14], and zygomycete fungi can be used as a substitute for fish feed [15]. These are filamentous fungi that contain large amounts of polyunsaturated fatty acids and protein, resulting in an increased content of these components in the fish feed [16]. Ferreira et al. (2016) [17] reported that ascomycete biomass (which has protein and fatty acid compositions comparable to zygomycete) is a good substitute for soybean-based feeds in the diet of animals like poultry, cattle, chicken and fish.

Among the zygomycete strains that have been investigated for fish feed, *R. oryzae* is a promising microorganism. *R. oryzae* has been used for many centuries in Asian cuisine to prepare fermented food, such as tempeh. Therefore, it is Generally Regarded As Safe (GRAS), which is a very favorable property when investigating its potential use as animal feed [18]. Due to these arguments, the possibility of using HTPL to cultivate *R. oryzae* is studied in this work. Alternatively, HTPL can be used in anaerobic digestion (AD) to produce biogas. AD is considered to be a sustainable form of treating industrial waste while simultaneously producing energy in the form of biogas [2]. Production of biogas from waste can have several benefits, including reduction in the costs of waste treatment, contribution to global energy needs using relatively cheap feedstock, and lower environmental impact than conventional types of energy [19].

In the present study, techno-economic and life cycle assessments of the treatment and use of HTPL are performed for three scenarios. The current strategy of PPL production is compared with two alternative scenarios where the HTPL is used for (i) the cultivation of filamentous fungus *R. oryzae* biomass or (ii) the production of biogas. The data used in the analyses are obtained from experiment, the literature and industrial potato starch production plants.

2. Materials and Methods

2.1. Process Description

This study is based on a typical plant producing potato starch that operates for a period of six months per year. It processes 300,000 metric tons of potato, produces 62,000 metric tons of starch and discards 210,000 m^3 of PL per six month period. The pH of the liquor is adjusted to 5.3 before the dissolved proteins are coagulated by injecting steam with a temperature of 140 °C. The HTPL (i.e., the remaining liquid after protein removal) was characterized by Fang et al. (2011) [2].

2.1.1. Concentration of HTPL by Evaporation (Evaporation Scenario)

HTPL is evaporated to produce steam that is used at the same facility. After protein coagulation, approximately 84,848 metric tons of HTPL (per six month operational period) containing 3.3% (*w/w*) TS is sent to a boiler to be concentrated to PPL containing 40% (*w/w*) TS. The PPL was characterized by Souza Filho et al. (2017) [1]. Evaporation occurs at 122 °C and 2 bar and at a flow rate of 19.64 ton/h of HTPL, producing 18.1 ton/h of steam and 1.5 ton/h of PPL. Part of the steam (15%) is used to preheat the HTPL and the remaining part is used for other operations at the same facility. The process flow diagram (PFD) of this scenario is presented in Figure 1.

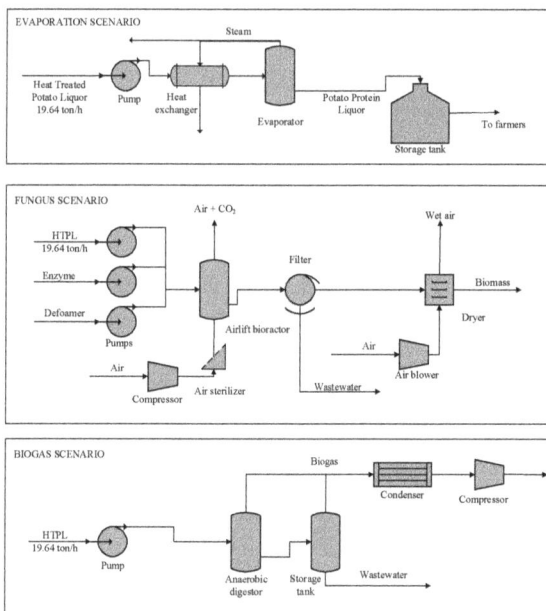

Figure 1. Process flow diagram of the three scenarios studied in this work.

2.1.2. Cultivation of Filamentous Fungus in HTPL (Fungus Scenario)

Experimental studies by Souza Filho et al. (2017) [1] indicate that *R. oryzae* grow best in the PPL waste stream when it is diluted back to 1:9 (i.e., 1 volume of PPL to 9 volumes of water). Therefore, the stream before concentration (i.e., the HTPL stream) is used in this modelling to cultivate *R. oryzae* under aerobic conditions using an airlift bioreactor. This type of bioreactor uses air, which is sparged in the medium, as the sole source of agitation. The aeration rate used in the reactor was calculated to keep the same gas holdup used by Souza Filho et al. (2017) [1]. The terminal velocity of a spherical bubble (v_t) in a bubble column is given by the Stokes' law:

$$v_t = \frac{g \cdot d^2 \cdot (\rho_l - \rho_g)}{18 \cdot \mu_l},$$

(1)

where g is the gravitational acceleration, d is the diameter of the bubble, ρ_l is the density of the medium, ρ_g is the density of the bubble and μ_l is the dynamic viscosity of the medium. The time it takes for a bubble to leave the liquid (Δt) is:

$$\Delta t = \frac{H}{v_t}, \tag{2}$$

where H is the height of the liquid column. During this time the volume of gas injected into the reactor (V_{air}) is:

$$V_{air} = Q_{air} \cdot \Delta t, \tag{3}$$

where Q_{air} is the air flow rate. The gas holdup (ε), defined as the volume of air divided by the volume of liquid present in the reactor, is:

$$\varepsilon = \frac{V_{air}}{V_{liquid}} = \frac{Q_{air} \cdot \Delta t}{V_{liquid}} = \frac{Q_{air} \cdot H}{v_t \cdot V_{liquid}}, \tag{4}$$

Assuming that the properties of the liquid and gas phases are the same in the experimental work and in the simulation (i.e., v_t is constant), and that the gas holdup is the same in experiment and in the simulation, then the air flow rate in the simulated reactor can be calculated from the experimental data using:

$$Q_{air2} = \frac{Q_{air1} \cdot H_1}{V_{liquid1}} \cdot \frac{V_{liquid2}}{H_2}, \tag{5}$$

The subscript 2 represents the properties in the simulated reactor and the subscript 1 the properties in the reactor used in the previous work. The height:diameter ratio of the simulated reactor was kept the same as the one used in the experimental work by Souza Filho et al. (2017) [1].

The proteins present in the HTPL induce the formation of foam [7]. Therefore, 0.2% (*v*/*v* of HTPL) defoamer is used in the reactor. Moreover, invertase is added at a proportion of 32.6 U per g of HTPL to assist the hydrolysis of the sucrose in the medium. After cultivation, the broth is sent to filters to separate the fermented broth from the fungal biomass. The broth containing low COD is sent to a wastewater treatment plant. The collected biomass is dried using hot air and used as fish feed. The operational conditions used in this simulation are presented in Table 1.

Table 1. Technical values used for the Fungus Scenario [1].

Type	Assumption
Reactor type	Airlift
Dilution rate	0.1 h^{-1}
Temperature	35 °C
Biomass yield	4.6 g/L HTPL
Nitrogen content in biomass	7.456% (*w*/*w*)

[1] Data based on [1].

2.1.3. Anaerobic Digestion of HTPL for Biogas Production (Biogas Scenario)

Fang et al. (2011) [2] have investigated the production of biogas from HTPL using an expanded granular sludge bed (EGSB) reactor. They found that the bioreactor can be operated continuously at a hydraulic retention time of 8 days, removing 87% of the COD in the form of biogas. The specifications that were used to simulate the biogas digester are presented in Table 2. The digestate resulting from the biogas production was kept in a storage tank for 20 h to remove the residual methane dissolved before being sent to a wastewater treatment plant. It was assumed that the biogas that is produced is used directly, without upgrading, in a combined heat and power (CHP) plant in the vicinity of the facility that produces the potato starch. The biogas is compressed to 5 bar before being sent to the gas grid.

Table 2. Technical values used for simulation of the fungus scenario [1].

Type	Assumption
Reactor type	EGSB
Hydraulic retention time (HRT)	8 days
Organic loading rate (OLR)	3.2 g COD/Lreactor.day
Temperature	37 °C
Methane production rate	1420 mL CH_4/Lreactor.day
Methane concentration	58% (v/v)
VFA content in the bioreactor	1 mM
Biogas pressure in the distribution pipeline	5 bar

[1] Data based on [2].

2.2. Energy, Equipment, and Economic Analyses

The energy, equipment and economic aspects were studied using Aspen Plus® V9 (Aspentech, Burlington, MA, USA) integrated with Aspen Energy Analyzer. The simulated data was exported to the Aspen Process Economic Analyzer software, where economic assumptions were entered. All economic assumptions used in this study are listed in Table 3. A modified version of the activity coefficient model NRTL (i.e., ELECNRTL) was used in all of the scenarios to include the effect of the electrolytes present in the HTPL. All of the equipment was made of stainless steel or carbon steel. The simulations included the purchase of one back-up pump identical to the original pump for all pumps. No further sterilization of the HTPL was considered in the fungus scenario, since the HTPL passes a heat treatment process in the starch plant. Contamination risks were not considered during the economic evaluation of the Fungus Scenario.

Table 3. Economic evaluation inputs and operational cost.

Type	Assumption
Annual operating time	4368 h (26 weeks)
Depreciation method	Straight line
Working capital [1]	15%/period
Tax rate [1]	33%/period
Interest rate [1]	6%/period
Lifetime of the plant [1]	20 years
Salvage value [1]	20% of initial capital cost
Operator labor	20 €/h
Supervisor labor	35 €/h
Electricity [1]	0.0775 €/kW·h
Steam [1]	0.01 €/kg
Wastewater treatment [1]	0.001 €/m³
Fish meal [2]	0.929 €/kg
Digestible crude protein content in fish meal [3]	65.6% (DM)
Digestible crude protein content in fungal biomass [4]	44.1% (DM)
Price conversion rate fungal biomass/fish meal	0.672
Invertase [5]	2.25×10^{-5} €/U
Defoamer [6]	2.3 €/L
Biogas [7]	33 €/MW·h
Low heat value Biogas (58% CH_4)	5.47 kW·h/Nm³

[1] Data based on [20]; [2] [21]; [3] [22]; [4] [16]; [5] [23]; [6] [24]; [7] [25]. DM: dry matter.

The price of fungal biomass was estimated using the price of fish meal adjusted by a factor based on the digestible protein content of both materials (see Table 3). The biogas price was calculated according to the market price of biogas in Sweden and the low heat value of the produced biogas. The revenue from the steam produced in the evaporation scenario is calculated from the regular steam price (see Table 3), even though it is used in the same plant. Economic calculations using the Aspen

Process Economic Analyzer (Aspentech, Burlington, MA, USA) were performed based on the prices from the first quarter of 2015. Capital costs, operating costs, product sales and net present value (NPV) were calculated considering a lifetime of the plant of 20 years.

2.3. Life Cycle Assessment (LCA)

This study uses a Consequential Life Cycle Assessment (CLCA) approach. CLCA assesses the environmental impact of products and yields information regarding the consequences as a result of marginal changes [26]. Therefore, it includes activities that are directly or indirectly affected by a marginal change in the level of output of a product [27]. Within the CLCA approach, system expansion is used to handle coproducts. In this method, the boundaries of the system are expanded to include the environmental impacts of alternative processes that produce the same products or functions as the studied coproducts [28]. The main product in this study is the supply of a treatment service, i.e., the service of treating the HTPL for further use. The coproducts are considered to substitute products that are already available on the market [29]. The products that are substituted are shown in Table 4.

Table 4. Coproducts obtained from the three scenarios studied here and the alternative products that are replaced by the coproducts.

Scenario	Coproduct	Replaced Product for Coproduct
Evaporation	PPL	Fertilizers [1]
Fungus	Fungal biomass	Fishmeal [2]
Biogas	Electricity	Marginal market for electricity in Sweden [3]
	Heat	Biomass in CHP plant [3]

[1] Marginal fertilizers: Calcium ammonium nitrate and potassium chloride [30]. Inventory data for fertilizer production retrieved from consequential life cycle assessment (CLCA) EcoInvent database [31]; [2] Data from CLCA EcoInvent database [31]; [3] Inventory data for fishmeal production retrieved from Fréon et al. [32].

2.3.1. System Boundaries, Functional Unit, and Environmental Impact Categories

The system boundaries for the three scenarios are show in Figure 2. It is assumed that the geographical boundaries for the systems are within Sweden. The functional unit is the treatment of one ton of HTPL residue. It is assumed that the waste material enters the system burden free, i.e., without any environmental impact associated with it. The selected environmental impact categories were global warming potential, acidification, fresh water ecotoxicity, as well as terrestrial, marine and freshwater eutrophication. The life cycle impact assessment (LCIA) methodologies recommended by the International Reference Life Cycle Data System (ILCD) were used [33]. The selection of the impact categories allows comparison of the systems for different environmental burdens and geographical scales (global and regional), in order to be able to identify and avoid solutions that could decrease local impacts but increase global burdens or vice versa [34]. The calculations were done using SimaPro v.8.3 (PRé Sustainability: Amersfoort, The Netherlands).

2.3.2. Basic Assumptions and Data Sources

Fungal Biomass

It was assumed that the fungal biomass is used to replace conventional fish meal in the market [35]. Table 5 shows the values for energy and protein content of the biomass and the fish meal. The substitution rate of the fish meal by the fungal biomass used the same factor previously mentioned for the economic analysis and that is shown in Table 3.

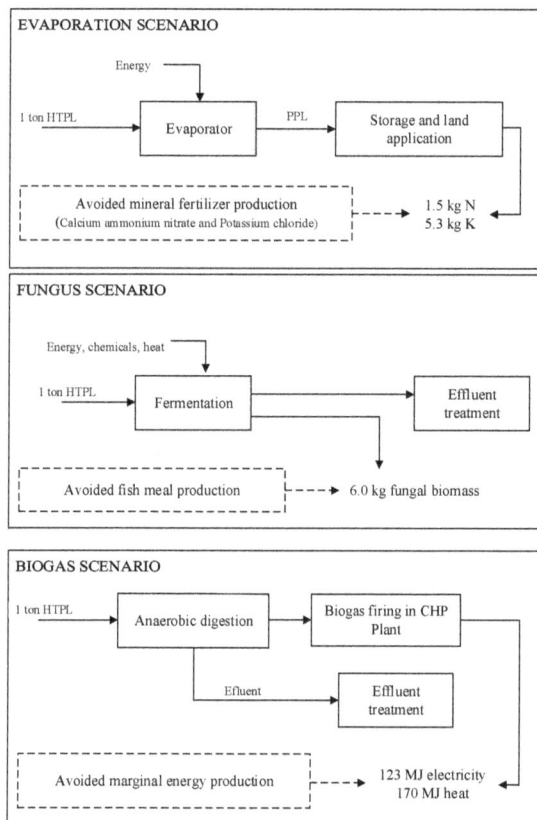

Figure 2. System boundaries and reference flows for the evaporation, fungus and biogas scenarios. The dotted lines show the avoided products in the system expansion.

Table 5. Biochemical composition of the fungal biomass and the substituted products.

Biochemical Parameter	Unit	Fish Meal [1]	Fungal Biomass [2]
Gross energy	MJ/kg DM	20.4	20.2
Digestible energy	MJ/kg DM	16.7 [3]	16.34
Crude protein	% DM	70.6	47.5

[1] [22]; [2] [16]; [3] Salmonid digestible energy.

Treatment of Wastewater

The effluent after the fungi cultivation (fungus scenario) or the AD (biogas scenario) requires further treatment. The wastewater treatment was adapted from the process "Wastewater from potato starch production" from EcoInvent Consequential database [31] based on the wastewater composition from Souza Filho et al. (2017) [1] and Fang et al. (2011) [2] for the fungus and biogas scenarios, respectively.

Nutrient Recovery

The PPL in the evaporation scenario is used as organic fertilizer for nitrogen and potassium. It was assumed that this organic fertilizer substitutes—and hence avoids the production of—the mineral fertilizers calcium ammonium nitrate and potassium chloride [30]. The emissions to air and water when using nitrogen for fertilizer on land used data from Tonini, Hamelin [30], which are average values in the literature regarding emissions to air and water from organic residues used as fertilizer [36–38].

Transportation of Coproducts

The PPL produced in the evaporation scenario is collected by farmers and transported an average of 100 km. The fungal biomass produced in the fungus scenario was considered to be transported 300 km to be used as feed in aquaculture production in western Sweden, where 10% of the national fish production occurs [39].

3. Results and Discussion

Production of potato starch generates a protein-rich side stream which is exploited by the industry to produce protein isolate. The residual wastewater is given (without charge) to the farmers to be used as fertilizer. This scenario was compared to other scenarios in which the wastewater is used to produce fungal biomass for use as fish feed or to produce biogas.

3.1. Technical Analysis

In the evaporation scenario, 19,641 kg/h of HTPL are pumped to a boiler operating at 2 bar to concentrate the HTPL to PPL, which contains approximately 40% (w/w) solids. The boiler produces 18,126 kg/h of steam which is used to preheat the HTPL before it enters the boiler. Approximately 267 MW·h/day of energy are consumed in this process.

The fungus scenario, involving the production of *R. oryzae* biomass to be used as fish feed, was evaluated for the same flow rate used for the evaporation scenario (19,641 kg/h of HTPL during six months a year). The cultivation of *R. oryzae* yielded a biomass production of 2475 kg/day (445 metric tons for an operational period of six months) containing 46.6% crude protein. The energy consumption in this process is 24.5 MW·h/day, which is primarily due to the aeration of the bioreactor (95% of the energy consumed in the scenario is for aeration). A daily volume of 473.8 m^3 of wastewater is discarded by the plant. An airlift bioreactor was chosen because of the improved agitation achieved in this design without the use of internal parts (e.g., impellers and baffles) in which the fungus can grow around interfering in the mass transfer [40].

In the biogas scenario, 279.1 Nm^3/h of biogas containing 58.8% (v/v) of methane is produced. At this concentration, the biogas contains a low heat value of 5.47 kW·h/Nm^3. The biogas production is equivalent to 36.6 MW·h/day, while the energy consumption is 11.9 MW·h/day, i.e., less than half of the energy needed for the fungus scenario. This is because the anaerobic digester uses mechanical agitation to create homogeneous conditions inside the reactor, as opposed to the airlift bioreactor used in the fungus scenario, which uses aeration as the source of agitation. This decreases the energy demand in the biogas scenario. 474.0 m^3 of wastewater is generated each day. Compared to the evaporation scenario, which is presently the most common alternative in potato starch plants, both bioprocess scenarios reduce the energy consumption.

3.2. Economic Analysis

Treatment of HTPL to PPL in the evaporation scenario requires an operating cost of €1.7 million per operating period (6 months). No income when using the PPL as fertilizer was accounted for in this scenario, since PPL is given without charge to the farmers for use as fertilizer. The excess steam produced and not used to preheat the HTPL represents an income of €671,414. The capital cost for this scenario is approximately €16.5 million. The evaporation of HTPL, containing as little as 3.3% (w/w) of TS, which is required to obtain the highly-concentrated PPL, demands the highest amount of heat when compared to the other scenarios. A standard vertical vessel was used to estimate the cost of the boiler for direct steam injection. Also, a shell and tube heat exchanger was designed to preheat the

HTPL. The equipment, size and construction material used in the simulation, as well as the individual prices, are presented in Table 6.

Table 6. Equipment costs for the different scenarios.

Scenario	Equipment	Capacity/Size [1]	Material	Cost (thousand €)
Evaporation	HTPL pump	6.2 L/s	SS	8.2
	Heat exchanger	114 m^2	SS	61.5
	Evaporator	650 m^2	CS	5450.5
	Storage tank	7000 m^3	CS	107.8
Fungus	HTPL pump	6.2 L/s	SS	8.2
	Air compressor for fermenter	467 m^3/h	CS	153.0
	Sterile air filter	467 m^3/h	[2]	2.0
	Pump for defoamer	4.4 mL/s	CS	4.1
	Pump for enzyme	98 mL/s	CS	4.1
	Airlift fermenter	200 m^3	SS	1478.5
	Biomass filter	9.3 m^2	CS	109.0
	Biomass dryer	9.3 m^2	CS	47.5
	Air blower for dryer	3095 m^3/h	CS	10.9
Biogas	HTPL pump	6.2 L/s	SS	8.2
	EGSB digester	4800 m^3	CS	3952.2
	Storage tank	480 m^3	CS	29.7
	Water condenser	1.6 m^2	SS	10.3
	Biogas compressor	260 m^3/h	SS	879.9

[1] Heat exchangers and filters defined by the surface area. Pumps and compressors defined by the flow rate. [2] Filter material cannot be adjusted in Aspen Process Economic Analyzer.

Production of fungal biomass (fungus scenario) demands much less energy. Only 9.2% of the energy presently used in the evaporation scenario would be required to produce the fungal biomass from HTPL. Cultivating the fungus on the potato starch wastewater has a capital cost of about €7.5 million. The cost of the airlift bioreactor was considered to be the same as the cost of a jacketed vertical tank. A rotary drum filter was used to collect the biomass after fermentation. Drying the biomass was achieved using a direct contact rotary dryer. The compressor used to provide air to the bioreactor was designed to provide an aeration of 0.04 vvm, and was calculated using Equation (4). The costs of the equipment are presented in Table 6. The operational cost of the plant was estimated to be €1.84 million per operating period, and the fungal biomass obtained during this period was sold as fish feed supplement for €282,150.

The biogas scenario has the lowest energy consumption (11.9 MW·h/day). The capital investment for this scenario is about €14.2 million. Operational costs are €1.4 million/period (including sending wastewater to a municipal treatment plant), and €216,785/period would be obtained from selling the biogas. Compared to the fungus scenario, the biogas scenario demands 89% more capital investment and the operational cost is 24% lower. The capital cost, operating cost and product sales for the proposed scenarios are presented in Figure 3. The digestate from the AD still contains nutrients which can be recovered in the form of fertilizer. However, the low concentration of such components in the digestate would require processes that have high energy demands, e.g., evaporation or centrifugation, or the transportation of large volumes of liquid. This would increase the costs associated with biogas scenario. Therefore, the wastewater produced in fungus and biogas scenarios is sent to the municipal wastewater treatment plant.

The NPV diagram after 10, 15 and 20 years is presented in Figure 4. No scenario returns the investment made. Fungal cultivation (fungus scenario) results in a NPV that is less negative than AD (biogas scenario). After 15 years, the NPV of the biogas scenario becomes similar to the NPV of the evaporation scenario and, at the end of the lifetime of the plant, evaporation and fungus scenarios have comparable NPV. This is caused by the large capital cost and low operational cost of the evaporation scenario, opposed to the low capital cost and large operational cost of the fungus scenario. The contrasting characteristics of the scenarios lead to a shift in the NPV during the last five years of the plant's lifetime. Rajendran et al. [20] estimated that the capital cost for a municipal solid waste (MSW) AD plant is about 40 million USD with operating costs of about 3 million USD per year. The plant was designed to treat 55,000 m^3 of MSW and to produce compressed biogas (CBG) for the transport sector. The MSW, which has a high TS content, yields 64 Nm^3 of raw biogas per m^3 of MSW versus 14.2 Nm^3 of raw biogas per m^3 of HTPL. This creates a situation where a plant can make a profit from waste treatment. In the case of the HTPL, the dilute nature of the waste stream requires larger equipment and higher energy consumption, and returns lower quantities of biogas, hence making it difficult to operate the treatment processes with a positive economic balance.

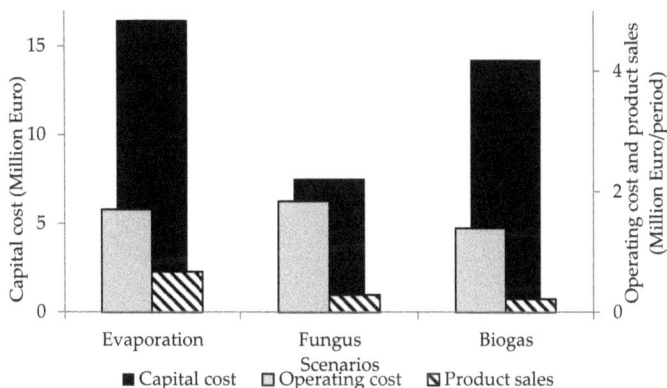

Figure 3. Results from the economic evaluation for the different scenarios considered in this study. The period is one year with six operational months.

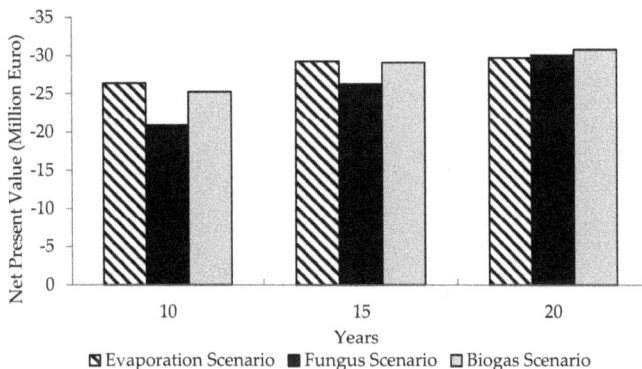

Figure 4. Net present value for the different scenarios after 10, 15 and 20 years.

3.3. Life Cycle Assessment

Figure 5 shows the environmental impacts of the three scenarios for HTPL treatment and use. The results show that the evaporation scenario has the largest impact in all of the impact categories except freshwater ecotoxicity. This is primarily due to the impacts related to the large amount of heat required for the evaporation of HTPL to PPL, which is part of the process emissions seen in the figure. Due to this heat requirement, the evaporation scenario has a large environmental impact despite the abatement from the avoided production of mineral fertilizer (seen as nutrients recovery in Figure 5).

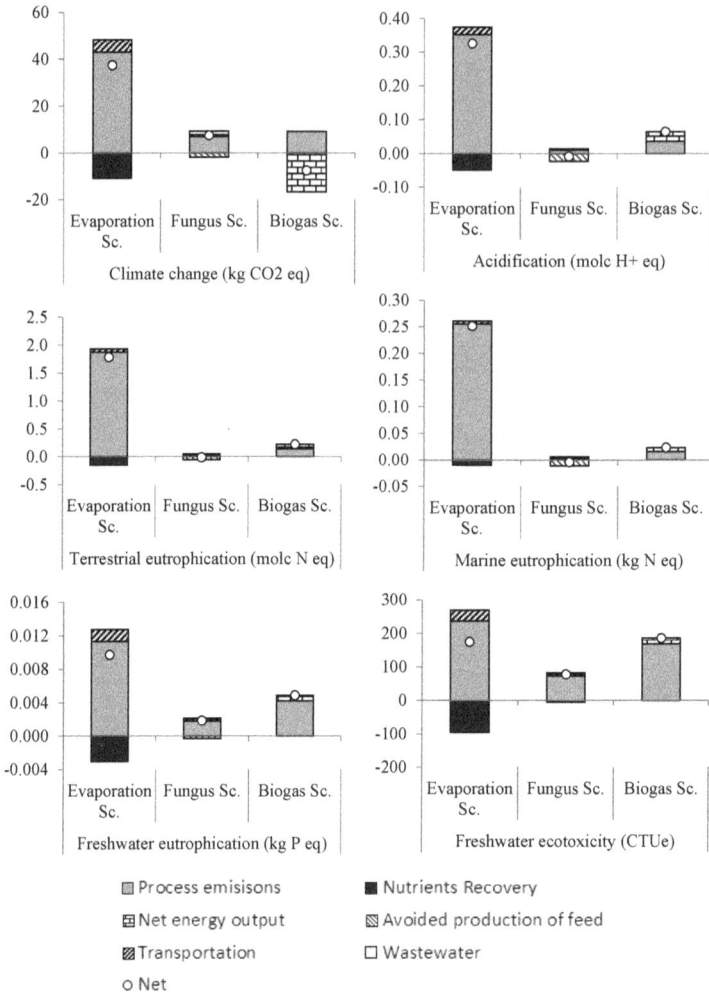

Figure 5. Environmental impacts of the three scenarios studied in this work.

The fungus scenario has lower impacts than the evaporation scenario in all of the impact categories. It also has a lower impact than the biogas scenario in all of the impact categories except climate change. The lower impact of the biogas scenario on climate change is mainly due to the avoided marginal energy production, which is a result from the biogas firing in a CHP plant. It can also be noted that biogas scenario has lower impacts than the evaporation scenario in all impact categories except freshwater ecotoxicity.

The trends seen in the acidification, terrestrial eutrophication, marine eutrophication and freshwater eutrophication impact categories are the same, with the fungus scenario having the lowest impact and the evaporation scenario the highest. The impact of fungus scenario on freshwater eutrophication is 77% and 55% lower compared to the evaporation and biogas scenarios, respectively (Figure 5). The results show that for the freshwater ecotoxicity category, the fungus scenario has the best performance with impacts that are 48% and 51% lower compared to the evaporation and biogas scenarios, respectively (Figure 5).

3.4. Comparison of the Different Scenarios

The fungus scenario has the lowest impact in five out of the six environmental impact categories that were analyzed. This may indicate that it is the preferred option. However, it must be emphasized that this scenario has a larger impact than the biogas scenario in the climate change category. Since this impact category is considered by the United Nations "the single biggest threat to development" [41], this result may have a central role when selecting the preferred scenario.

The fungus scenario is also the preferred scenario according to the economic analysis, since it has the lowest capital cost and the best NPV during the first fifteen years of operation. In contrast, at the end of the plant's lifetime, the evaporation scenario becomes economically more viable. However, the difference between the two scenarios is only 1% of the evaporation scenario's NPV.

The evaporation scenario has the largest impact in five out of the six environmental impact categories, in addition to having the largest capital investment of all scenarios.

Since none of the scenarios was best in all of the analyzed parameters, it is not possible to draw a simple conclusion regarding the preferred scenario. A decision would ultimately be made according to the political, environmental or economic agenda of the decision-makers and identifying the crucial factors can be difficult [42]. The most important contribution of this study is to highlight the trade-offs inherently involved in the decision process.

4. Conclusions

Technical, economic and environmental analyses were performed to determine potential benefits of two proposed scenarios to a plant discarding 19.64 ton/h of HTPL. The two proposed scenarios are to use the HTPL (i) to cultivate filamentous fungus *R. oryzae* to produce a protein-rich biomass (fungus scenario) and (ii) to produce biogas via AD (biogas scenario). These two scenarios are compared to the most commonly used treatment method, which is concentrating the HTPL before using it as a fertilizer. Both proposed scenarios reduce the capital cost and the energy consumption of the wastewater treatment. Moreover, the current study highlights the environmental benefits of cultivating fungi in the HTPL (fungus scenario), since it has the lowest impact in acidification, freshwater ecotoxicity as well as the terrestrial, freshwater, and marine eutrophication categories. In contrast, the greenhouse gas emissions were higher from fungus scenario compared to biogas scenario, where the residue was anaerobically digested. The results show that the substituted products in the system expansion, such as mineral fertilizers, electricity and heat, substantially reduce the environmental footprints of fungus and biogas Scenarios. This study presents the techno-economic and environmental trade-offs that are necessary to take into account when selecting one of the scenarios in preference to the others.

Acknowledgments: The authors would like to acknowledge the Coordination for the Improvement of Higher Education Personel (CAPES-Brazil), and the Gunnar Ivarsson foundation for financing this work.

Author Contributions: Pedro F Souza Filho, Akram Zamani, and Mohammad J. Taherzadeh conceived and designed the simulation scenarios. Pedro Brancoli and Kim Bolton were responsible for the life cycle analysis. Pedro F Souza Filho and Pedro Brancoli wrote the paper. Kim Bolton, Akram Zamani, and Mohammad J. Taherzadeh reviewed and edited the manuscript.

Conflicts of Interest: The authors declare no conflict of interest.

Abbreviations

AD	Anaerobic digestion
CBG	Compressed biogas
CHP	Combined heat and power
CLCA	Consequential life cycle assessment
COD	Chemical oxygen demand
CS	Carbon steel
DM	Dry matter
EGSB	Expanded granular sludge bed

EU	European Union
GRAS	Generally regarded as safe
HRT	Hydraulic retention time
HTPL	Heat treated potato liquor
LCA	Life cycle assessment
LCIA	Life cycle impact assessment
MSW	Municipal solid waste
NPV	Net present value
OLR	Organic loading rate
PFD	Process flow diagram
PL	Potato liquor
PP	Potato pulp
PPL	Potato protein liquor
SS	Stainless steel
TS	Total solids
U	Unity of enzyme activity
USD	United States dollar
VFA	Volatile fatty acid

References

1. Souza Filho, P.F.; Zamani, A.; Taherzadeh, M.J. Production of edible fungi from potato protein liquor (PPL) in airlift bioreactor. *Fermentation* **2017**, *3*, 12. [CrossRef]
2. Fang, C.; Boe, K.; Angelidaki, I. Biogas production from potato-juice, a by-product from potato-starch processing, in upflow anaerobic sludge blanket (UASB) and expanded granular sludge bed (EGSB) reactors. *Bioresour. Technol.* **2011**, *102*, 5734–5741. [CrossRef] [PubMed]
3. Ralet, M.-C.; Guéguen, J. Fractionation of potato proteins: Solubility, thermal coagulation and emulsifying properties. *LWT-Food Sci. Technol.* **2000**, *33*, 380–387. [CrossRef]
4. Zhang, D.-Q.; Mu, T.-H.; Sun, H.-N.; Chen, J.-W.; Zhang, M. Comparative study of potato protein concentrates extracted using ammonium sulfate and isoelectric precipitation. *Int. J. Food Prop.* **2017**, *20*, 2113–2127. [CrossRef]
5. Waglay, A.; Karboune, S.; Alli, I. Potato protein isolates: Recovery and characterization of their properties. *Food Chem.* **2014**, *142*, 373–382. [CrossRef] [PubMed]
6. Strætkvern, K.O.; Schwarz, J.G. Recovery of native potato protein comparing expanded bed adsorption and ultrafiltration. *Food Bioprocess Technol.* **2012**, *5*, 1939–1949. [CrossRef]
7. Bárta, J.; Heřmanová, V.; Diviš, J. Effect of low-molecular additives on precipitation of potato fruit juice proteins under different temperature regimes. *J. Food Process Eng.* **2008**, *31*, 533–547. [CrossRef]
8. Klingspohn, U.; Bader, J.; Kruse, B.; Kishore, P.V.; Schuegerl, K.; Kracke-Helm, H.A.; Likidis, Z. Utilization of potato pulp from potato starch processing. *Process Biochem.* **1993**, *28*, 91–98. [CrossRef]
9. Zwijnenberg, H.J.; Kemperman, A.J.B.; Boerrigter, M.E.; Lotz, M.; Dijksterhuis, J.F.; Poulsen, P.E.; Koops, G.-H. Native protein recovery from potato fruit juice by ultrafiltration. *Desalination* **2002**, *144*, 331–334. [CrossRef]
10. Klingspohn, U.; Vijai Papsupuleti, P.; Schügerl, K. Production of enzymes from potato pulp using batch operation of a bioreactor. *J. Chem. Technol. Biotechnol.* **1993**, *58*, 19–25. [CrossRef]
11. Lotz, M.; Fröhlich, R.; Matthes, R.; Schügerl, K.; Seekamp, M. Bakers' yeast cultivation on by-products and wastes of potato and wheat starch production on a laboratory and pilot-plant scale. *Process Biochem.* **1991**, *26*, 301–311. [CrossRef]
12. Kumar, P.K.R.; Singh, A.; Schügerl, K. Fed-batch culture for the direct conversion of cellulosic substrates to acetic acid/ethanol by *Fusarium oxysporum*. *Process Biochem.* **1991**, *26*, 209–216. [CrossRef]
13. Schügerl, K.; Rosen, W. Investigation of the use of agricultural byproducts for fungal protein production. *Process Biochem.* **1997**, *32*, 705–714. [CrossRef]
14. Olli, J.J.; Krogdahl, Å.; van den Ingh, T.S.; Brattås, L.E. Nutritive value of four soybean products in diets for atlantic salmon (*Salmo salar*, L.). *Acta Agric. Scand. Sect. A* **1994**, *44*, 50–60. [CrossRef]

15. Ferreira, J.A.; Lennartsson, P.R.; Niklasson, C.; Lundin, M.; Edebo, L.; Taherzadeh, M.J. Spent sulphite liquor for cultivation of an edible *Rhizopus* sp. *BioResources* **2012**, *7*, 173–188.
16. Ferreira, J.A.; Lennartsson, P.R.; Taherzadeh, M.J. Production of ethanol and biomass from thin stillage by *Neurospora intermedia*: A pilot study for process diversification. *Eng. Life Sci.* **2015**, *15*, 751–759. [CrossRef]
17. Ferreira, J.A.; Mahboubi, A.; Lennartsson, P.R.; Taherzadeh, M.J. Waste biorefineries using filamentous ascomycetes fungi: Present status and future prospects. *Bioresour. Technol.* **2016**, *215*, 334–345. [CrossRef] [PubMed]
18. Ferreira, J.A.; Lennartsson, P.R.; Edebo, L.; Taherzadeh, M.J. Zygomycetes-based biorefinery: Present status and future prospects. *Bioresour. Technol.* **2013**, *135*, 523–532. [CrossRef] [PubMed]
19. Mao, C.; Feng, Y.; Wang, X.; Ren, G. Review on research achievements of biogas from anaerobic digestion. *Renew. Sustain. Energy Rev.* **2015**, *45*, 540–555. [CrossRef]
20. Rajendran, K.; Kankanala, H.R.; Martinsson, R.; Taherzadeh, M.J. Uncertainty over techno-economic potentials of biogas from municipal solid waste (MSW): A case study on an industrial process. *Appl. Energy* **2014**, *125*, 84–92. [CrossRef]
21. Indexmundi. Available online: http://www.indexmundi.com/commodities/?commodity=fish-meal (accessed on 12 September 2017).
22. Feedipedia. Available online: https://www.feedipedia.org/node/208 (accessed on 12 September 2017).
23. Invertase on alibaba.com. Available online: https://www.alibaba.com/product-detail/INVERTASE_60590660279.html (accessed on 12 September 2017).
24. Defoamer for Fermentation on alibaba.com. Available online: https://www.alibaba.com/product-detail/Defoamer-for-Fermentation_1473403819.html (accessed on 12 September 2017).
25. Joelsson, E.; Dienes, D.; Kovacs, K.; Galbe, M.; Wallberg, O. Combined production of biogas and ethanol at high solids loading from wheat straw impregnated with acetic acid: Experimental study and techno-economic evaluation. *Sustain. Chem. Processes* **2016**, *4*, 14. [CrossRef]
26. Ekvall, T.; Weidema, B.P. System boundaries and input data in consequential life cycle inventory analysis. *Int. J. Life Cycle Assess.* **2004**, *9*, 161–171. [CrossRef]
27. Brander, M.; Tipper, R.; Hutchison, C.; Davis, G. *Technical Paper: Consequential and Attributional Approaches to LCA: A Guide to Policy Makers with Specific Reference to Greenhouse Gas LCA of Biofuels*; Ecometrica Press: London, UK, 2009.
28. Ekvall, T.; Finnveden, G. Allocation in ISO 14041—A critical review. *J. Clean. Prod.* **2001**, *9*, 197–208. [CrossRef]
29. By-products, Recycling and Waste. Available online: https://consequential-lca.org/clca/by-products-recycling-and-waste/ (accessed on 12 September 2017).
30. Tonini, D.; Hamelin, L.; Wenzel, H.; Astrup, T. Bioenergy production from perennial energy crops: A consequential LCA of 12 bioenergy scenarios including land use changes. *Environ. Sci. Technol.* **2012**, *46*, 13521–13530. [CrossRef] [PubMed]
31. Wernet, G.; Bauer, C.; Steubing, B.; Reinhard, J.; Moreno-Ruiz, E.; Weidema, B. The Ecoinvent database version 3 (part I): Overview and methodology. *Int. J. Life Cycle Assess.* **2016**, *21*, 1218–1230. [CrossRef]
32. Fréon, P.; Durand, H.; Avadí, A.; Huaranca, S.; Moreyra, R.O. Life cycle assessment of three peruvian fishmeal plants: Toward a cleaner production. *J. Clean. Prod.* **2017**, *145*, 50–63. [CrossRef]
33. European Commission. *International Reference Life Cycle Data System (ILCD) Handbook—General Guide for Life Cycle Assessment—Provisions and Action Steps*, 1st ed.; Publications Office of the European Union: Ispra, Italy, 2011; ISBN 978-92-79-17451-3.
34. Van der Werf, H.M.G.; Petit, J. Evaluation of the environmental impact of agriculture at the farm level: A comparison and analysis of 12 indicator-based methods. *Agric. Ecosyst. Environ.* **2002**, *93*, 131–145. [CrossRef]
35. Edebo, L.B. Zygomycetes for Fish Feed. Patent WO2008/002231, 3 January 2008.
36. De Klein, C.; Novoa, R.S.A.; Ogle, S.; Smith, K.A.; Rochette, P.; Wirth, T.C.; McConkey, B.G.; Mosier, A.; Rypdal, K.; Walsh, M.; et al. N_2O emissions from managed soils, and CO_2 emissions from lime and urea application. In *2006 IPCC Guidelines for National Greenhouse Gas Inventories*; Eggleston, H.S., Buendia, L., Miwa, K., Ngara, T., Tanabe, K., Eds.; IPCC National Greenhouse Gas Inventories Programme: Kanagawa, Japan, 2006; Volume 4, pp. 1–54, ISBN 4-88788-032-4.

37. Hamelin, L.; Jørgensen, U.; Petersen, B.M.; Olesen, J.E.; Wenzel, H. Modelling environmental consequences of direct land use changes from energy crops in a self-sustained and fully renewable energy system in Denmark: Effect of crop types, soil, climate, residues management, initial carbon level and turnover time. In Proceedings of the Quantifying and Managing Land Use Effects of Bioenergy, Campinas, Brazil, 19–21 September 2011.

38. Galloway, J.N.; Dentener, F.J.; Capone, D.G.; Boyer, E.W.; Howarth, R.W.; Seitzinger, S.P.; Asner, G.P.; Cleveland, C.C.; Green, P.A.; Holland, E.A.; et al. Nitrogen cycles: Past, present, and future. *Biogeochemistry* **2004**, *70*, 153–226. [CrossRef]

39. Statistiska Centralbyrån. *Aquaculture in Sweden in 2014*; Statistiska Centralbyrån: Stockholm, Sweden, 2014.

40. Ferreira, J.A.; Lennartsson, P.R.; Taherzadeh, M.J. Airlift bioreactors for fish feed fungal biomass production using edible filamentous fungi. In Proceedings of the FFBiotech Symposium, Villeneuve d'Ascq, France, 15–16 May 2017.

41. United Nations. Goal 13: Sustainable Development Knowledge Platform. Available online: https://sustainabledevelopment.un.org/sdg13 (accessed on 13 October 2017).

42. Eriksson, O.; Bisaillon, M.; Haraldsson, M.; Sundberg, J. Enhancement of biogas production from food waste and sewage sludge—Environmental and economic life cycle performance. *J. Environ. Manag.* **2016**, *175*, 33–39. [CrossRef] [PubMed]

fermentation

MDPI

Article

Tuning of the Carbon-to-Nitrogen Ratio for the Production of L-Arginine by *Escherichia coli*

Mireille Ginésy * , Daniela Rusanova-Naydenova and Ulrika Rova *

Biochemical Process Engineering, Division of Chemical Engineering, Department of Civil, Environmental and Natural Resources Engineering, Luleå University of Technology, SE-971 87 Luleå, Sweden; daniela.rusanova-naydenova@ltu.se
* Correspondence: mireille.ginesy@ltu.se (M.G.); ulrika.rova@ltu.se (U.R.); Tel.: +46-920-491-315 (U.R.)

Received: 6 October 2017; Accepted: 6 November 2017; Published: 10 November 2017

Abstract: L-arginine, an amino acid with a growing range of applications within the pharmaceutical, cosmetic, food, and agricultural industries, can be produced by microbial fermentation. Although it is the most nitrogen-rich amino acid, reports on the nitrogen supply for its fermentation are scarce. In this study, the nitrogen supply for the production of L-arginine by a genetically modified *Escherichia coli* strain was optimised in bioreactors. Different nitrogen sources were screened and ammonia solution, ammonium sulphate, ammonium phosphate dibasic, and ammonium chloride were the most favourable nitrogen sources for L-arginine synthesis. The key role of the C/N ratio for L-arginine production was demonstrated for the first time. The optimal C/N molar ratio to maximise L-arginine production while minimising nitrogen waste was found to be 6, yielding approximately 2.25 g/L of L-arginine from 15 g/L glucose with a productivity of around 0.11 g/L/h. Glucose and ammonium ion were simultaneously utilized, showing that this ratio provided a well-balanced equilibrium between carbon and nitrogen metabolisms.

Keywords: *Escherichia coli*; fermentation; L-arginine; carbon to nitrogen ratio; nitrogen sources

1. Introduction

L-arginine is a semi-essential amino acid commonly used in pharmaceutical, nutraceutical, and cosmetic industries [1]. It can also be used as animal feed or fertilizer [2]. *Corynebacterium glutamicum* and *C. crenatum* have been commonly used for the microbial production of L-arginine [3–6]. Environmental concerns prompt a sustainable use of raw materials that are widely available, easily renewable, and that do not compete with food production. One such feedstock is lignocellulosic biomass which mainly consists of cellulose, hemicellulose, and lignin [7]. In contrast to the cellulose fraction, the hemicellulose fraction can include several additional monosaccharides besides glucose, i.e., xylose, mannose, galactose, rhamnose, and arabinose, where the composition and structure vary depending on the species and origin of the lignocellulose source. Cost-efficient use of this feedstock requires a microorganism able to use both five and six carbon sugars. However, neither *C. glutamicum* nor *C. crenatum* strains naturally metabolize five-carbon sugars. On the other hand, *Escherichia coli* is able to use pentoses as well as hexoses for the fermentative production of several different molecules with an industrial value [8–10]. Combined with its fast growth, its robustness and the availability of molecular tools for its genetic engineering, *E. coli* is also a candidate for L-arginine production. An *E. coli* strain able to produce nearly 12 g/L of L-arginine with a yield of 0.17 $g_{arginine}/g_{glucose}$ was recently engineered [11].

L-arginine biosynthesis is a nitrogen-requiring process since this amino acid consists of 32% nitrogen. Moreover, nitrogen, required for protein synthesis, is a vital nutrient for cell growth. Nitrogen represents 14% of the cell dry mass in growing *E. coli* [12]. Sufficient nitrogen to support growth and high L-arginine production must therefore be provided during fermentation.

However, after a certain threshold, the addition of ammonium sulphate, a common nitrogen source, was detrimental for the production of lysine and succinate by *C. glutamicum* [13,14] and that of L-threonine and L-phenylalanine by *E. coli* [15,16]. In addition, an excessive supply of nitrogen might result in large nitrogen wastes, which pose serious environmental threats such as global warming, thinning of the stratospheric ozone layer, and biodiversity loss [17–21]. Despite the fact that the supply of nitrogen is clearly a key parameter in the fermentation process and must be finely adjusted, studies on nitrogen for L-arginine production are lacking.

It is well known that the carbon-to-nitrogen (C/N) ratio is a crucial parameter in some microbial processes, such as biogas production [22–24] and lipid production by oleaginous yeasts [25,26]. It has also been demonstrated that the C/N ratio has a great influence on growth and metabolite production for a variety of microorganisms [27,28].

Based on fermentation data, transcriptional RNA level, and enzyme activity, it has also been shown that the *E. coli* metabolism is affected by the C/N ratio [29]. The importance of optimizing the C/N ratio in *E. coli* fermentations has further been demonstrated for heterologous gene expression [30] and for the production of the amino acid L-threonine, where the C/N molar ratio resulting in the best production was 69 [15].

The carbon to nitrogen ratio is of such importance because the metabolisms of carbon and nitrogen are tightly linked where their assimilation is coordinated and controlled by the availability of both nutrients [29,31–33]. Indeed, *E. coli* possesses two pathways for nitrogen assimilation (Figure 1b,c) and both pathways require 2-oxoglutarate, one of the key intermediates of the TCA cycle, to convert ammonia into glutamate [34]. The C/N ratio might be of even greater significance for the biosynthesis of L-arginine which involves both acetyl-CoA (the starting point of the TCA cycle) and glutamate (Figure 1a).

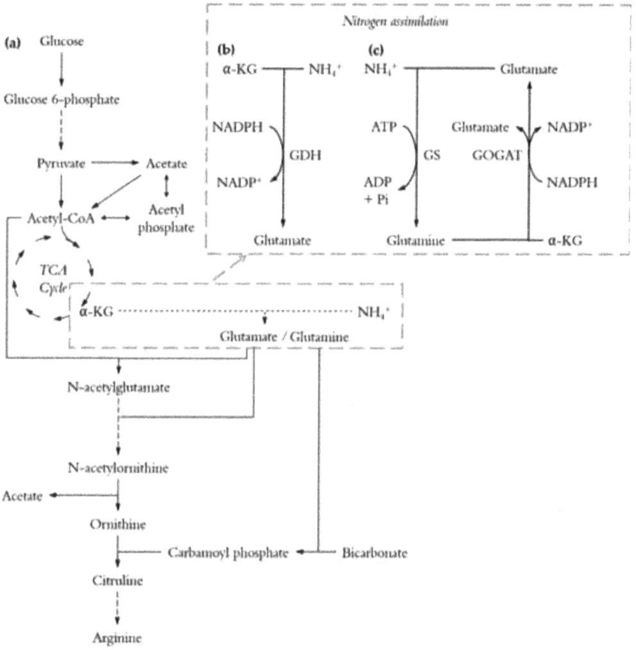

Figure 1. Simplified pathways for L-arginine biosynthesis. Long-dashed lines: multi-step reactions; short-dashed lines: several possible pathways; α-KG: α-ketoglutarate. (**a**) Pathway from glucose to L-arginine; (**b**) The glutamate dehydrogenase (GDH) pathway; (**c**) The glutamine synthetase-glutamate synthase (GS-GOGAT) pathway.

As a substantial amount of nitrogen is required for the fermentative production of L-arginine, high-cost organic nitrogen sources, such as yeast extract or tryptone, would not be suitable for large-scale production. Only enterohemorrhagic *E. coli* strains have been shown to grow on urea, a low-cost organic nitrogen source, whereas other strains have not displayed any urease activity [35,36].

In this work, various inorganic nitrogen sources were screened and the most suitable ones were used to investigate, for the first time, the effect of the C/N ratio on the production of L-arginine. Batch fermentations in minimal medium were performed in bioreactors using a genetically modified *E. coli* strain.

2. Materials and Methods

2.1. Microorganism

The strain *E. coli* SJB009 previously engineered to have an enhanced L-arginine production ability was used [11]. This strain is derived from *E. coli* K-12 C600. Genes responsible for L-arginine catabolism (*adiA*, *speC* and *speF*) and for the repression of the L-arginine biosynthesis genes (*argR*) were knocked-out. In *E. coli*, the first enzyme of the dedicated pathway for L-arginine biosynthesis (encoded by *argA*) is sensitive to feedback inhibition by L-arginine. In the strain SJB009, a variant (*argA214*) coding for a feedback resistant enzyme was introduced and overexpressed, and the wild type *argA* was deleted. Finally, the L-arginine export gene (*argO*) was overexpressed. Stock cultures of the strain were stored at $-80\,^\circ$C in 15% glycerol.

2.2. Seed Cultures for Fermentations

Shake flasks containing 100 mL Luria Bertani (LB) medium were sterilized at 121 $^\circ$C for 20 min and then inoculated with 500 μL stock cultures and incubated at 32 $^\circ$C and 200 rpm for 12 h. The cells were harvested by centrifugation at 4 $^\circ$C and 5000 rpm for 10 min. The cells were then washed twice by resuspension in phosphate-buffered saline solution (pH 7) and centrifugation under the same conditions. Subsequently, the cells were resuspended in 25 mL sterilised fermentation medium described below and aseptically inoculated into the bioreactors.

2.3. Fermentations

Duplicate batch fermentations were performed in 1 L bioreactors (Biobundle 1 L, Applikon Biotechnology) with a working volume of 700 mL. In order to avoid interactions with nitrogen contained in complex media, the cultivations were carried out in a defined medium consisting of (per liter): 3 g KH_2PO_4, 12.8 g $Na_2HPO_4 \cdot 7H_2O$, 0.5 g NaCl, 1 g $MgSO_4 \cdot 7H_2O$, 20 mg $FeSO_4 \cdot 7H_2O*$, 12 mg $MnSO_4 \cdot 7H_2O*$, 1 mg $CaCl_2*$, 0.25 g antifoam, and 20 mg tetracycline·HCl* (*: added after sterilisation at 121 $^\circ$C for 20 min). The reactors contained either 30 or 15 g/L of glucose and an appropriate amount of the nitrogen source to obtain the desired C/N molar ratio. They represent the ratios of the molar concentration of carbon (from glucose) to that of nitrogen.

Prior to the fermentation, the pH was adjusted to 7 using 2 M HCl or 5 M NaOH when necessary. The temperature was set to 32 $^\circ$C and the stirring speed to 500 rpm [11]. Throughout the fermentation, the pH was maintained at 7 with automatic addition of 5 M NaOH and the dissolved oxygen level at 50% with automatic addition of 5 vvm of air. Samples were regularly taken for cell growth, glucose, ammonium ion, L-arginine, and acetic acid measurements.

In the first step, fermentations were performed with an initial glucose concentration of 15 g/L to screen for potential nitrogen sources for L-arginine production by *E. coli*. Seven inorganic nitrogen sources were compared using a C/N ratio of 6: ammonia solution (NH_4OH), ammonium carbonate (($NH_4)_2CO_3$), ammonium chloride (NH_4Cl), ammonium nitrate (NH_4NO_3), ammonium phosphate dibasic (($NH_4)_2HPO_4$), ammonium sulphate (($NH_4)_2SO_4$), and sodium nitrate ($NaNO_3$). The carbon atom from the carbonate group of ammonium carbonate was not taken into consideration to calculate the C/N ratio. In addition, one organic nitrogen source, monosodium glutamate, was assessed at three different concentrations (2.5, 5, and 10 g/L).

In the second step, the most efficient nitrogen sources were used to compare three C/N molar ratios (3, 6, and 12) at two initial glucose concentrations (15 and 30 g/L).

2.4. Cell Growth Analyses

The dry cell weight (DCW) was determined by washing cells contained in 5 mL fermentation broth and measuring their weight after drying in a furnace at 80 °C for 24 h.

2.5. Substrates, Products and By-Products Analyses

The samples were centrifuged for 10 min at 4 °C and $10,600 \times g$ and the supernatant was filtered through 0.2 μm filters.

Glucose, L-arginine, and glutamate were analysed with the Dionex AAA-Direct™ system (Thermo Scientific, Waltham, MA, USA): a high-pressure ion chromatography system (Dionex-ICS 5000⁺ HPIC, Thermo Scientific) equipped with an electrochemical detector (ICS-5000⁺ ED, Thermo Scientific), an anion exchange column (AminoPac PA10 Analytical Column, Thermo Scientific), and a guard column (AminoPac PA10 Guard Column, Thermo Scientific). Analyses were performed at ambient conditions, with deionized water and 250 mM NaOH as eluents (for glutamate 1 M sodium acetate was also used) and a flow rate of 0.75 mL/min.

Acetic acid, the main by-product formed during fermentation [11], was quantified by HPLC using refractive index detection. The cation exchange column (Aminex HPX87-H, BioRad, Hercules, CA, USA) and guard column (Micro-Guard IG Cation H Cartridge, BioRad) were maintained at 65 °C. The mobile phase was 5 mM H_2SO_4 at a flow rate of 0.6 mL/min.

Ammonium ion concentration (NH_4^+) was determined using a colorimetric test based on the Nessler method (MQuant™, Merck, Darmstadt, Germany). The samples were diluted to contain no more than 400 mg/L ammonium and subsequently analysed.

3. Results and Discussion

3.1. Comparison of Nitrogen Sources

The effect of different inorganic nitrogen sources on *E. coli* growth and L-arginine production was investigated with an initial glucose concentration of 15 g/L and a C/N ratio of 6. Monosodium glutamate was also tested as nitrogen source at various concentrations. The results obtained with all sources studied are summarized in Table 1.

Table 1. Summary of fermentation results for the different nitrogen sources. Duplicate bioreactor cultivations in minimal medium, 15 g/L initial glucose, 32 °C, 500 rpm, air at 5 vvm, pH 7. Ammonium salts and ammonia solution were provided so as to have a C/N ratio of 6. Results are given as means ± standard deviations.

Nitrogen Source	μ (1/h)	DCW (g/L)	L-arginine (g/L)	Q_p (g/L/h)
Ammonia solution	0.10 ± 0.01	2.35 ± 0.08	2.30 ± 0.09	0.12 ± 0.01
Ammonium carbonate	0.06 *	2.08 ± 0.02	2.11 ± 0.01	0.08 *
Ammonium chloride	0.09 *	2.18 ± 0.03	2.23 ± 0.06	0.12 ± 0.01
Ammonium nitrate	0.06 *	2.31 ± 0.12	1.44 ± 0.11	0.05 *
Ammonium phosphate dibasic	0.09 *	2.25 ± 0.03	2.25 ± 0.06	0.13 ± 0.02
Ammonium sulphate	0.09 ± 0.01	2.30 ± 0.10	2.41 ± 0.08	0.13 ± 0.01
Sodium nitrate	nd	nd	nd	nd
Monosodium glutamate 10 g/L	nd	nd	nd	nd
Monosodium glutamate 5 g/L	nd	nd	nd	nd
Monosodium glutamate 2.5 g/L	0.01 *	0.52 ± 0.02	nd	nd

* standard deviation was lower than 0.01. nd: not detected. μ: specific growth rate; DCW: dry cell weight; Q_p: volumetric productivity.

Under those conditions, ammonia solution, ammonium chloride, ammonium phosphate dibasic, and ammonium sulphate all yielded a bit over 2.2 g/L of L-arginine with a productivity of about 0.1 g/L/h. These nitrogen sources also resulted in similar cell growth, in terms of both maximum dry cell weight and specific growth rate.

The use of ammonium carbonate also resulted in similar L-arginine production (2.11 g/L), but the cells grew and produced L-arginine more slowly (0.06 h^{-1} and 0.08 g/L/h, respectively).

Neither growth nor L-arginine production occurred during fermentation with sodium nitrate, demonstrating the need for an ammonium-containing nitrogen source. This was expected since it has been determined that *E. coli* cannot normally utilize nitrate aerobically [37]; although cell growth and xylanase production have been reported for a different *E. coli* strain [30].

Accordingly, only the nitrogen from the ammonium group was taken into account when adjusting the C/N ratio with ammonium nitrate. With this nitrogen source, the final cell density (2.31 g/L) was comparable to what was achieved with other ammonium salts; however, the growth rate was a bit lower (0.06 h^{-1}), less L-arginine was produced (1.44 g/L), and the productivity was much lower (0.05 g/L/h).

As glutamate is the main precursor to L-arginine biosynthesis (Figure 1a), it was reasoned that a supply of nitrogen directly in this form might boost L-arginine production. The sodium glutamate was used at three different concentrations: 10 g/L, 5 g/L, and 2.5 g/L; with a glucose concentration of 15 g/L. After five days, no growth was observed in the fermentations with 10 g/L and 5 g/L monosodium glutamate. At a monosodium glutamate concentration of 2.5 g/L, only very weak growth occurred. In the three cases, no L-arginine production was detected, showing that glutamate is not suitable as the sole nitrogen source. Poor growth of a variety of *E. coli* strains on glucose when glutamate (at about 1.5 g/L) was the sole nitrogen source has been reported in a recent study [38]. It has been observed that under those conditions, TCA cycle intermediates accumulate, leading to the inhibition of the signalling molecule cAMP levels, thereby impairing cell growth. This accumulation was suggested to be caused by an imbalance between carbon and nitrogen due to a slow glutamate uptake combined with a rapid glucose consumption.

3.2. Comparison of C/N Ratios

The C/N ratio is an important parameter in many microbial processes. In order to determine its influence on L-arginine production by *E. coli*, three different ratios were tested: 3, 6, and 12. Each ratio was studied with an initial glucose concentration of 30 g/L and 15 g/L, respectively, to check whether the results obtained were solely due to the C/N ratio or to the changes in nitrogen concentration.

The four inorganic nitrogen sources that were found to be the most effective for this fermentation were used: ammonium sulphate, ammonium phosphate dibasic, ammonium chloride, and ammonia solution. Using different nitrogen sources for each ratio enabled a thorough comparison of those sources and ensured that the observed effects were dependent on the amount of ammonium rather than on one particular counterion.

The DCW and L-arginine concentration for every source, each taken when the highest L-arginine concentration was reached, are presented in Figure 2. In addition, the productivity, the glucose consumption rate, and the L-arginine yields from glucose and ammonium are provided in supporting information.

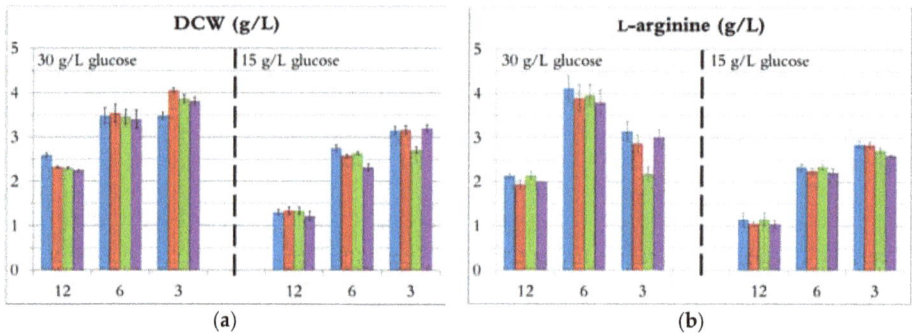

Figure 2. The influence of C/N ratio on fermentation. $(NH_4)_2SO_4$ in blue; $(NH_4)_2HPO_4$ in red; NH_4Cl in green; NH_4OH in purple.—(**a**) DCW; (**b**) L-arginine production.

Ammonia solution, ammonium chloride, ammonium dibasic phosphate, and ammonium sulphate were nearly equally suitable for L-arginine production by *E. coli* in minimal medium. Indeed, at each ratio, the different fermentation results were relatively similar for each nitrogen source (Figure 2), with the exception of ammonium chloride showing somewhat lower productivities. For any given C/N ratio, the fermentation profiles were almost identical regardless of the nitrogen source. The curves of cell growth, glucose, L-arginine, and acetate concentration during fermentations with ammonium sulphate are shown as examples in Figure 3.

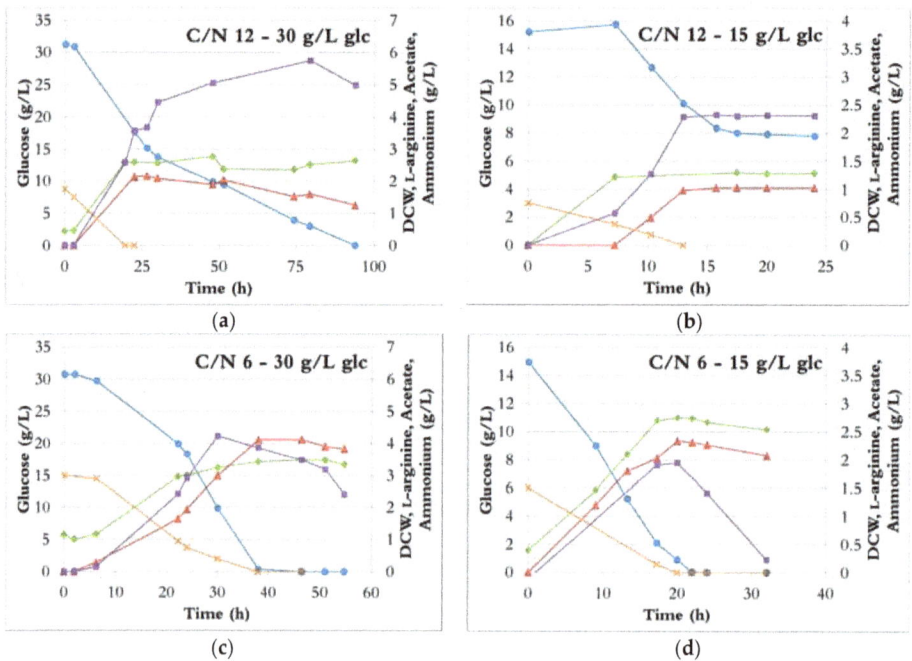

Figure 3. *Cont.*

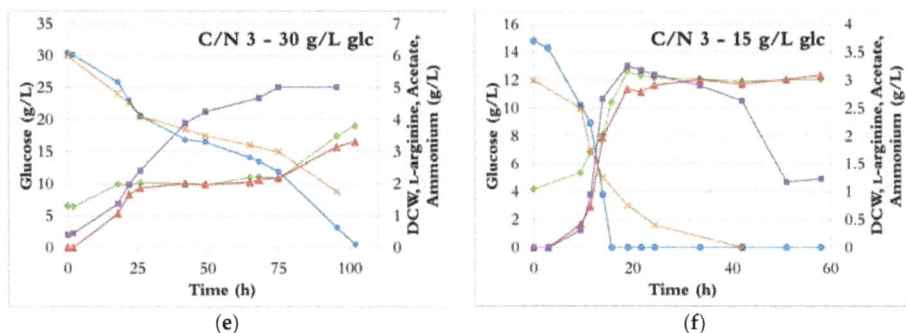

Figure 3. The fermentation profiles with different C/N ratios during growth in bioreactors with ammonium sulphate as the sole nitrogen source. Blue circle: glucose; green tilted square: dry cell weight; red triangle: L-arginine; purple square: acetate; orange cross: ammonium. (**a**) C/N ratio of 12, 30 g/L glucose; (**b**) C/N ratio of 12, 15 g/L glucose; (**c**) C/N ratio of 6, 30 g/L glucose; (**d**) C/N ratio of 6, 15 g/L glucose; (**e**) C/N ratio of 3, 30 g/L glucose; (**f**) C/N ratio of 3, 15 g/L glucose.

At a C/N ratio of 12 (Figure 3a,b), ammonium was depleted in less than 20 h and 14 h when the initial glucose concentration was 30 g/L and 15 g/L, respectively. Both cell growth and L-arginine production stopped with the depletion of ammonia, although about half of the glucose was still unused. With both 30 g/L and 15 g/L initial glucose, the DCW (around 2.4 g/L and 1.3 g/L, respectively) and the L-arginine concentrations (around 2 g/L and 1 g/L, respectively) were the lowest obtained (Figure 2). These results therefore indicate that a C/N ratio of 12 provides insufficient nitrogen for efficient fermentation. In the fermentation with 30 g/L glucose, the cells used the remaining glucose after cell growth and L-arginine production stopped, likely due to utilization of L-arginine as a poor nitrogen source (L-arginine concentration is decreasing) [38], and only acetate was produced. In the fermentation with 15 g/L glucose, neither glucose uptake nor acetate production occurred after ammonia depletion. It can be speculated that a longer time might have been needed for the cells to adapt to the new conditions (no ammonia available), possibly because the cell density was very low (1.3 g/L).

At a C/N ratio of 6, ammonium and glucose were nearly simultaneously depleted after 40 h (30 g/L initial glucose, Figure 3b) and 20 h (15 g/L initial glucose, Figure 3d). The DCW increased by either 50% (30 g/L initial glucose) or 100% (15 g/L initial glucose) compared to a C/N ratio of 12 (Figures 2 and 3a). In both cases, L-arginine production almost doubled and reached 3.9 ± 0.1 g/L and 2.3 ± 0.1 g/L, from 30 and 15 g/L glucose, respectively (Figure 2b). Hence, a C/N ratio of 6 seemed to be suitable for L-arginine production.

At a C/N ratio of 3, with an initial glucose concentration of 15 g/L, the glucose was rapidly depleted (16–24 h), while a third of the ammonium was still available (Figure 3f). After glucose exhaustion, the cells slowly used acetate and gradually consumed all the ammonium, albeit at a decreasing rate, but neither growth nor L-arginine production occurred. Even though the ammonium provided was doubled compared to a C/N ratio of 6, the L-arginine production (2.8 ± 0.1 g/L) was only increased by about 25%; the L-arginine yields from ammonium were thus significantly lower (Figure S1b). However, the productivities were increased by 30 to 60% depending on the nitrogen source (Figure S2b).

With the same ratio but 30 g/L initial glucose, cells stopped growing and producing L-arginine after approximately 25 h, although about two thirds of the glucose and of the ammonium were left in the broth (Figure 3e). Glucose and ammonium consumption rates dropped while acetate was still rapidly produced. Cells resumed a normal behaviour after a 50 h lag; at that time, the ammonium concentration was down to 3 g/L and the glucose concentration to about 12 g/L. The glucose was the limiting nutrient and was depleted in 25 h, whereas 25% of the ammonium was still unused. Cell growth and L-arginine production were halted. In the end, the cells had grown to similar or higher densities than at a C/N ratio of 6 (Figure 2a), but needed three times longer to reach it (e.g., Figure 3c,e). On the other hand, the L-arginine production (2.8 ± 0.4 g/L) was at least 25% lower than at a C/N ratio of 6. Compared to a C/N of 6 or 12, the volumetric productivities and the glucose consumption rates dropped by 80% (Figure S2). The lowest L-arginine yields from glucose (0.11 ± 0.01 mol/mol) and ammonium (0.07 ± 0.01 mol/mol) were obtained (Figure S1).

Consequently, a C/N ratio of 3 resulted in a waste of ammonium, not only because much was yet unused when the highest L-arginine concentration was reached, but also because the rest was not efficiently utilised for L-arginine production.

It is not clear why the cells started to grow and produce L-arginine normally and then experienced a long lag phase before resuming normal activity. However, the fact that fermentation was hampered with 30 g/L initial glucose but not with 15 g/L suggests that it is the high concentration of ammonium, or the combination of the high concentrations of ammonium and glucose, rather than the C/N ratio itself, that had such a detrimental effect. A negative impact of excessive ammonium on fermentation has been observed in other studies [15,16,39]. For instance, increasing the initial ammonium sulphate concentration from 10 g/L to 20 g/L resulted in a nearly six-fold decline in L-phenylalanine production and productivity [16]. A further increase of the ammonium sulphate concentration (30 g/L) caused a 17-fold drop of these parameters compared to 10 g/L of ammonium sulphate. Here, similar results were obtained since the production and productivity of L-arginine decreased five-fold when increasing the ammonium sulphate from 13.2 g/L to 26.4 g/L (C/N ratio of 6 and 3, respectively, 30 g/L initial glucose).

In most fermentations, the cells were able to use the acetate they had previously excreted as the sole carbon source (Figure 3). However, this merely permitted cell survival and supported neither cell growth nor L-arginine production.

Looking more closely into the acetate production is interesting since acetate is the main by-product during L-arginine production [11]. Conditions that minimise its formation are desirable. Indeed acetate can be toxic for cells [40] and its production requires carbon that might otherwise be used for L-arginine production. Acetate formation cannot be completely avoided since L-arginine biosynthesis requires the split of N-acetylornithine into ornithine and acetate (Figure 1a). However, this only accounts for 1 mole of acetate per mole of L-arginine and most of the acetate is produced from acetyl-CoA or pyruvate (Figure 1a).

Regardless of the C/N ratio, around 4 g/L were produced with 30 g/L initial glucose and 2 to 3 g/L with 15 g/L initial glucose. However, the mechanism behind acetate formation cannot be the same at all C/N ratios. Indeed, half as much L-arginine is produced with a ratio of C/N equal to 12 compare to one of 6 (Figure 2b). Therefore, half as much of the total acetate comes from the split of N-acetylornithine (Figure S3) and the production of acetate relative to that of L-arginine is twice as high (Figure S3c). Moreover, the acetate yields from glucose with a C/N ratio of 12 are about twice as high than those from a C/N ratio of 6 (Figure S3d).

Together, these results show that at a C/N ratio of 12 a majority of glucose is diverted toward acetate formation via acetyl-CoA and pyruvate. This could be because less α-ketoglutarate is needed for nitrogen assimilation, leading to a more active TCA cycle and thus more NADH production. Too much NADH results in an increase in pyruvate concentration and inhibition of the TCA cycle. Acetate formation represents an alternative for carbon utilization resulting in less NADH formation [41].

A C/N ratio of 6 resulted in the lowest ratios of acetate to L-arginine (around 3 mol/mol), which confirms that this ratio provides a good balance between the carbon and the nitrogen metabolism.

For a C/N ratio of 3, with 15 g/L glucose, the acetate yield from glucose, the acetate to L-arginine ratio, and the acetate formed from the L-arginine pathway are sensibly similar to those with a C/N of 6 (Figure S3b–d). However, with 30 g/L, more acetate is formed from acetyl-CoA and/or pyruvate with a C/N ratio of 3 than with a C/N ratio of 6, although the acetate yields are similar or even slightly lower for a C/N of 3. Once again, this shows that a high ammonium concentration in that case has a negative impact on the fermentation.

An initial C/N ratio of 6 was the most suitable for this fermentation, resulting in good growth and L-arginine production with reasonably low acetate formation. It should be noted that a lower initial glucose concentration leads to higher L-arginine yields from both glucose (0.17 ± 0.01 vs. 0.14 ± 0.01 mol/mol) and ammonium (0.17 ± 0.02 vs. 0.13 mol/mol) (Figure S1).

4. Conclusions

Ammonia solution, ammonium sulphate, and ammonium phosphate dibasic were the most effective nitrogen sources for L-arginine production. Ammonium chloride was nearly as good but resulted in slightly lower productivities. The ammonium carbonate yielded a reasonable L-arginine concentration but at a significantly lower productivity. The other nitrogen sources tested gave either poor L-arginine production (ammonium nitrate) or no production (sodium nitrate, monosodium glutamate).

This study highlights the limitations of a batch process for the large scale microbial production of L-arginine. Indeed, with 15 g/L glucose, a C/N ratio of 3 resulted in the highest productivities; however, maintaining this ratio for a higher glucose concentration is not feasible as it implies an excessive ammonium source concentration, as observed here with 30 g/L glucose.

With a C/N ratio of 3 and 15 g/L glucose, only two thirds of the ammonium was consumed by the time glucose was exhausted. In a fed-batch or continuous process, a C/N of 3 in the inflowing medium would therefore most likely result in the accumulation of the nitrogen source to toxic levels. A C/N ratio of 12 is not recommended for these processes as it favors glucose conversion to acetate rather than to L-arginine.

A C/N ratio of 6, however, provided an excellent balance between the carbon and the nitrogen metabolism during batch fermentations. Both the glucose and the ammonium source were efficiently used for L-arginine production and simultaneously depleted. This suggests that this ratio would be well suited for either fed-batch or continuous fermentations.

In addition, it was shown that using a lower initial glucose concentration (15 g/L) resulted in higher L-arginine yields from glucose and ammonium and reduced acetate formation.

Under the optimal conditions (C/N ratio of 6, initial glucose of 15 g/L), about 2.29 g/L of L-arginine was formed at a productivity of 0.11 g/L/h; L-arginine yields were 0.17 mol/mol from both glucose and ammonium.

Supplementary Materials: The following are available online at www.mdpi.com/2311-5637/3/4/60/s1.

Acknowledgments: The authors would like to thank Kempestiftelserna and Bio4Energy, a strategic research environment appointed by the Swedish government, for supporting this work.

Author Contributions: Mireille Ginésy conceived, designed, and conducted the experiments, analysed the data, and wrote the paper. Daniela Rusanova-Naydenova contributed to samples analysis by ion chromatography and revised the manuscript. Ulrika Rova participated in the experimental design and revised the manuscript.

Conflicts of Interest: The authors declare no conflict of interest.

References

1. Loscalzo, J. L-arginine and atherothrombosis. *J. Nutr.* **2004**, *134*, 2798S–2800S; discussion 2818S–2819S. [PubMed]
2. Öhlund, J.; Näsholm, T. Low nitrogen losses with a new source of nitrogen for cultivation of conifer seedlings. *Environ. Sci. Technol.* **2002**, *36*, 4854–4859. [CrossRef] [PubMed]
3. Chen, M.; Chen, X.; Wan, F.; Zhang, B.; Chen, J.; Xiong, Y. Effect of Tween 40 and DtsR1 on L-arginine overproduction in *Corynebacterium crenatum*. *Microb. Cell Fact.* **2015**, *14*, 119. [CrossRef] [PubMed]
4. Nakayama, K.; Yoshida, H. Fermentative production of L-arginine. *Agric. Biol. Chem.* **1972**, *36*, 1675–1684. [CrossRef]

5. Dou, W.; Xu, M.; Cai, D.; Zhang, X.; Rao, Z.; Xu, Z. Improvement of L-arginine production by overexpression of a bifunctional ornithine acetyltransferase in *Corynebacterium crenatum*. *Appl. Biochem. Biotechnol.* **2011**, *165*, 845–855. [CrossRef] [PubMed]
6. Park, S.H.; Kim, H.U.; Kim, T.Y.; Park, J.S.; Kim, S.; Lee, S.Y. Metabolic engineering of *Corynebacterium glutamicum* for L-arginine production. *Nat. Commun.* **2014**, *5*, 4618. [CrossRef] [PubMed]
7. Anwar, Z.; Gulfraz, M.; Irshad, M. Agro-industrial lignocellulosic biomass a key to unlock the future bio-energy: A brief review. *J. Radiat. Res. Appl. Sci.* **2014**, *7*, 163–173. [CrossRef]
8. Ghosh, D.; Hallenbeck, P.C. Fermentative hydrogen yields from different sugars by batch cultures of metabolically engineered *Escherichia coli* DJT135. *Int. J. Hydrog. Energy* **2009**, *34*, 7979–7982. [CrossRef]
9. Dien, B.S.; Nichols, N.N.; Bothast, R.J. Recombinant *Escherichia coli* engineered for production of L-lactic acid from hexose and pentose sugars. *J. Ind. Microbiol. Biotechnol.* **2001**, *27*, 259–264. [CrossRef] [PubMed]
10. Andersson, C.; Hodge, D.; Berglund, K.A.; Rova, U. Effect of different carbon sources on the production of succinic acid using metabolically engineered *Escherichia coli*. *Biotechnol. Prog.* **2007**, *23*, 381–388. [CrossRef] [PubMed]
11. Ginesy, M.; Belotserkovsky, J.; Enman, J.; Isaksson, L.; Rova, U. Metabolic engineering of *Escherichia coli* for enhanced arginine biosynthesis. *Microb. Cell Fact.* **2015**, *14*, 29. [CrossRef] [PubMed]
12. Neidhardt, F.C.; Ingraham, J.L.; Low, K.B.; Magasanik, B.; Schaechter, M.; Umbarger, H. *Escherichia coli and Salmonella Typhimurium: Cellular and Molecular Biology*; American Society for Microbiology: Washington, DC, USA, 1987; Volume 2.
13. Haleem Shah, A.; Hameed, A.; Ahmad, S.; Majid Khan, G. Optimization of culture conditions for L-lysine fermentation by *Corynebacterium glutamicum*. *J. Biol. Sci.* **2002**, *2*, 151–156.
14. Jeon, J.M.; Rajesh, T.; Song, E.; Lee, H.W.; Lee, H.W.; Yang, Y.H. Media optimization of *Corynebacterium glutamicum* for succinate production under oxygen-deprived condition. *J. Microbiol. Biotechnol.* **2013**, *23*, 211–217. [CrossRef] [PubMed]
15. Chen, N.; Huang, J.; Feng, Z.B.; Yu, L.; Xu, Q.Y.; Wen, T.Y. Optimization of fermentation conditions for the biosynthesis of L-threonine by *Escherichia coli*. *Appl. Biochem. Biotechnol.* **2009**, *158*, 595–604. [CrossRef] [PubMed]
16. Yuan, P.; Cao, W.; Wang, Z.; Chen, K.; Li, Y.; Ouyang, P. Enhancement of L-phenylalanine production by engineered *Escherichia coli* using phased exponential L-tyrosine feeding combined with nitrogen source optimization. *J. Biosci. Bioeng.* **2015**, *120*, 36–40. [CrossRef] [PubMed]
17. Hahn, J.; Crutzen, P.J. The role of fixed nitrogen in atmospheric photochemistry. *Philos. Trans. R. Soc. Lond. Ser. B* **1982**, *296*, 521–541. [CrossRef]
18. Kinzig, A.P.; Socolow, R.H. Human impacts on the nitrogen cycle. *Physics Today* **1994**, *47*, 24–31. [CrossRef]
19. Sutton, M.A.; Howard, C.M.; Erisman, J.W.; Billen, G.; Bleeker, A.; Grennfelt, P.; van Grinsven, H.; Grizzetti, B. *The European Nitrogen Assessment: Sources, Effects and Policy Perspectives*; Cambridge University Press: Cambridge, UK, 2011.
20. Galloway, J.N. The global nitrogen cycle: Changes and consequences. *Environ. Pollut.* **1998**, *102*, 15–24. [CrossRef]
21. Ramaswamy, V.; Boucher, O.; Haigh, J.; Hauglustaine, D.; Haywood, J.; Myhre, G.; Nakajima, T.; Shi, G.; Solomon, S.; Betts, R.E.; et al. *Radiative Forcing of Climate Change*; Houghton, J.T., Callander, B.A., Varney, S.K., Eds.; Cambridge University Press: New York, NY, USA, 2001.
22. Dioha, I.; Ikeme, C.; Nafi'u, T.; Soba, N. Effect of carbon to nitrogen ratio on biogas production. *IRJNS* **2013**, *1*, 1–10.
23. Wu, X.; Yao, W.; Zhu, J.; Miller, C. Biogas and CH$_4$ productivity by co-digesting swine manure with three crop residues as an external carbon source. *Bioresour. Technol.* **2010**, *101*, 4042–4047. [CrossRef] [PubMed]
24. Wang, X.; Yang, G.; Feng, Y.; Ren, G.; Han, X. Optimizing feeding composition and carbon–nitrogen ratios for improved methane yield during anaerobic co-digestion of dairy, chicken manure and wheat straw. *Bioresour. Technol.* **2012**, *120*, 78–83. [CrossRef] [PubMed]
25. Sattur, A.P.; Karanth, N.G. Production of microbial lipids: II. Influence of C/N ratio—model prediction. *Biotechnol. Bioeng.* **1989**, *34*, 868–871. [CrossRef] [PubMed]
26. Ykema, A.; Verbree, E.C.; Kater, M.M.; Smit, H. Optimization of lipid production in the oleaginous yeast *Apiotrichum curvatum* in wheypermeate. *Appl. Microbiol. Biotechnol.* **1988**, *29*, 211–218. [CrossRef]

27. Kalil, M.S.; Alshiyab, H.S.; Yusoff, W.M.W. Effect of nitrogen source and carbon to nitrogen ratio on hydrogen production using *C. acetobutylicum*. *Am. J. Biochem. Biotechnol.* **2008**, *4*, 393–401. [CrossRef]

28. Wang, D.; Wei, G.; Nie, M.; Chen, J. Effects of nitrogen source and carbon/nitrogen ratio on batch fermentation of glutathione by *Candida utilis*. *Korean J. Chem. Eng.* **2010**, *27*, 551–559. [CrossRef]

29. Kumar, R.; Shimizu, K. Metabolic regulation of *Escherichia coli* and its *gdhA, glnL, gltB* D mutants under different carbon and nitrogen limitations in the continuous culture. *Microb. Cell Fact.* **2010**, *9*, 8. [CrossRef] [PubMed]

30. Mohd Rusli, F.; Mohamed, M.S.; Mohamed, R.; Puspaningsih, N.N.T.; Ariff, A. Kinetics of xylanase fermentation by recombinant *Escherichia coli* DH5a in shake flask culture. *Am. J. Biochem. Biotechnol.* **2009**, *5*, 110–118.

31. Commichau, F.M.; Forchhammer, K.; Stülke, J. Regulatory links between carbon and nitrogen metabolism. *Microb. Cell Fact.* **2006**, *9*, 167–172. [CrossRef] [PubMed]

32. Mao, X.J.; Huo, Y.X.; Buck, M.; Kolb, A.; Wang, Y.P. Interplay between CRP-cAMP and PII-Ntr systems forms novel regulatory network between carbon metabolism and nitrogen assimilation in *Escherichia coli*. *Nucleic Acids Res.* **2007**, *35*, 1432–1440. [CrossRef] [PubMed]

33. Doucette, C.D.; Schwab, D.J.; Wingreen, N.S.; Rabinowitz, J.D. alpha-ketoglutarate coordinates carbon and nitrogen utilization via Enzyme I inhibition. *Nat. Chem. Biol.* **2011**, *7*, 894–901. [CrossRef] [PubMed]

34. Helling, R.B. Pathway choice in glutamate synthesis in *Escherichia coli*. *J. Bacteriol.* **1998**, *180*, 4571–4575. [PubMed]

35. Nakano, M.; Iida, T.; Ohnishi, M.; Kurokawa, K.; Takahashi, A.; Tsukamoto, T.; Yasunaga, T.; Hayashi, T.; Honda, T. Association of the urease gene with enterohemorrhagic *Escherichia coli* strains irrespective of their serogroups. *J. Clin. Microbiol.* **2001**, *39*, 4541–4543. [CrossRef] [PubMed]

36. Monk, J.M.; Charusanti, P.; Aziz, R.K.; Lerman, J.A.; Premyodhin, N.; Orth, J.D.; Feist, A.M.; Palsson, B.Ø. Genome-scale metabolic reconstructions of multiple *Escherichia coli* strains highlight strain-specific adaptations to nutritional environments. *Proc. Natl. Acad. Sci. USA* **2013**, *110*, 20338–20343. [CrossRef] [PubMed]

37. Kobayashi, M.; Ishimoto, M. Aerobic inhibition of nitrate assimilation in *Escherichia coli*. *Z. Allg. Mikrobiol.* **1973**, *13*, 405–413. [CrossRef] [PubMed]

38. Bren, A.; Park, J.O.; Towbin, B.D.; Dekel, E.; Rabinowitz, J.D.; Alon, U. Glucose becomes one of the worst carbon sources for *E. coli* on poor nitrogen sources due to suboptimal levels of cAMP. *Sci. Rep.* **2016**, *6*. [CrossRef] [PubMed]

39. Xu, F.; Gage, D.; Zhan, J. Efficient production of indigoidine in *Escherichia coli*. *J. Ind. Microbiol. Biotechnol.* **2015**, *42*, 1149–1155. [CrossRef] [PubMed]

40. Luli, G.W.; Strohl, W.R. Comparison of growth, acetate production, and acetate inhibition of *Escherichia coli* strains in batch and fed-batch fermentations. *Appl. Environ. Microbiol.* **1990**, *56*, 1004–1011. [PubMed]

41. Eiteman, M.A.; Altman, E. Overcoming acetate in *Escherichia coli* recombinant protein fermentations. *Trends Biotechnol.* **2006**, *24*, 530–536. [CrossRef] [PubMed]

fermentation

MDPI

Article

Green Biorefinery of Giant Miscanthus for Growing Microalgae and Biofuel Production

Shuangning Xiu *, Bo Zhang, Nana Abayie Boakye-Boaten and Abolghasem Shahbazi *

Department of Natural Resources and Environmental Design, North Carolina Agricultural and Technical State University, 1601 East Market Street, Greensboro, NC 27411, USA; bzhang@ncat.edu (B.Z.); naboakye@aggies.ncat.edu (N.A.B.-B.)
* Correspondence: xshuangn@ncat.edu (S.X.); ash@ncat.edu (A.S.); Tel.: +1-336-285-3830 (S.X.)

Received: 13 November 2017; Accepted: 7 December 2017; Published: 11 December 2017

Abstract: In this study, an innovative green biorefinery system was successfully developed to process the green biomass into multiple biofuels and bioproducts. In particular, fresh giant miscanthus was separated into a solid stream (press cake) and a liquid stream (press juice) using a screw press. The juice was used to cultivate microalga *Chlorella vulgaris*, which was further thermochemically converted via thermogravimetry analysis (TGA) and pyrolysis-gas chromatography/mass spectrometry (Py-GC/MS) analysis, resulting in an approximately 80% conversion. In addition, the solid cake of miscanthus was pretreated with dilute sulfuric acid and used as the feedstock for bioethanol production. The results showed that the miscanthus juice could be a highly nutritious source for microalgae that are a promising feedstock for biofuels. The highest cell density was observed in the 15% juice medium. Sugars released from the miscanthus cake were efficiently fermented to ethanol using *Saccharomyces cerevisiae* through a simultaneous saccharification and fermentation (SSF) process, with 88.4% of the theoretical yield.

Keywords: fermentation; ethanol production; green biorefinery; miscanthus; microalgae; thermochemical conversion

1. Introduction

Fossil fuels are the major source for our energy need at present, which currently contribute about 80% of the global energy demand [1]. According to the International Energy Agency (IEA), this demand will be increased by 40% by the year 2035, with fossil fuels contributing 75% [2]. This over dependence and increasing use of fossil fuels in the US and globally serves as a source of worry, as it is predicted to reach a crisis point in the near future [3].

Fossil fuels are also the major source of environmental pollutants and greenhouse gases. Sustainable developments requires the use of renewable biomass-based resources for fuels, chemicals and material production [4]. Renewable biomass resources can be converted to fuels and are a logical choice to replace oil. Considerable attention has been given to lignocellulosic biomass such as agricultural residues and energy crops for biofuel production.

In order to derive full benefits from the use of biomass in the biofuel industry, it is imperative that different byproducts are produced from the feedstock in a way analogous to the petroleum refinery platform. This concept of producing various bioproducts from a single biomass feedstock is known as the biorefinery concept. A biorefinery is a facility that can convert biomass into multiple biofuels, bioproducts, power, and chemicals by integrating various biomass conversion processes [1].

Currently, the biorefinery platform is aimed at replacing the petroleum refineries and reducing the fossil fuel intensity in different production areas [2]. As such, it is vital to investigate and develop biorefinery platforms based on lignocellulosic biomass and improve on their economic potentials to make them competitive with the petroleum refining industry.

One major type of the biorefinery concepts being explored is the green biorefinery. A green biorefinery uses "nature-wet" biomasses such as green grasses as feedstock, and employs a wet-fractionation technology as a first step to isolate the content-substances in their natural form into fiber-rich cake and a nutrient-rich green juice which are then converted by different processes into various products [5,6]. Using the nutrient-rich green juice for products generation eliminates the need for energy intensive drying processes, which could greatly improve the economic viability of this biorefinery platform type.

Using perennial energy grasses as feedstock for biorefinery processes has attracted tremendous attention due to their superior advantages such as broad adaptability, high water and fertilizer use efficiency, and tremendous biomass production. Miscanthus x. giganteus (MxG) is a perennial warm season grass in the sugarcane family that has been identified as a primary biomass crop for development in the US [7].

In order to enhance the versatility of the green biorefinery and increase the economics of the green biorefinery, it is essential to process both the press cake and the press juice into a variety of final products or other intermediary products that could be used as feedstock for further downstream processes [8]. The press cake has been primarily used for production of fodder pellet and biogas [9]. Little or no attention has yet been focused on the utilization of the press cake for biofuel purpose, especially as raw materials for bioethanol production. Bioethanol is by far the most widely used biofuel for transportation worldwide. Bioethanol and bioethanol/gasoline blends have a long history as alternative transportation fuels [10]. Producing bioethanol as a transportation fuel can help reduce CO_2 buildup in two important ways: by displacing the use of fossil fuels, and by recycling CO_2 that is released when it is combusted as fuel. The efficiency of conversion of biomass to ethanol depends upon feedstock characteristics and composition, pretreatment processes, and the fermentation technologies [11]. Research focus on the use of the press juice is on the increase now. Various uses of juice extracted from biomass have been investigated in the biorefinery platform. For example, the extracted banagrass juice has been used for the cultivation of an edible fungus [12]. Our previous study observed that the cattail juice held great potential as alternative growth media for microalgae and bacteria [13]. The juice from Italian rye grass, clover grass and alfalfa has been used for the production of lactic acids and other value added products [14].

In green biorefinery, wet fractionation technology is used as the first step to separate the juice from the green biomass. Screw press was the most common method used to press the green juice out of the green biomass [14]. Besides the screw press, filter presses, belt presses, centrifuges, Hammer mill, the thermal mechanical dewatering method, simultaneous application of a pulsed electric field, and superimposition of ultrasounds were also used to separate the juice from green biomass [15,16]. However, most of the publications did not provide a detailed description of the fractionation process, such as equipment operation, product recovery rates and analysis. Furthermore, to the best of our knowledge, there has not been any work done so far on the wet fractionation of MxG.

With the aim of overcoming the problems outlined above, the intent of this research is to develop a green biorefinery platform in an economically viable and environmentally sustainable manner, using MxG as a feedstock. Particularly, we separated MxG into pressed cake and juice, and accessed the viability of using the juice to grow microalgae for hydrocarbons production, and using the press cake to produce bioethanol using some biochemical processes. To the best of our knowledge, this use of juice from MxG has not been investigated yet.

2. Materials and Methods

2.1. Miscanthus Harvest and Processing

MxG was harvested from North Carolina A&T State University farm during the early summer in 2015 using a Tanaka TPH 270s-pole hedge trimmer to achieve consistent cuts. The MxG was then shred into smaller size using a DR Wood chipper/shredder (14.50 Pro Manual Start, DR Power Equipment,

Vergennes, VT, USA). A Carver press (#2094 Cage Equipment, Carver Inc., Wabash, IN, USA) was used to press and separate the shredded biomass into a green juice liquid fraction and a solid cake fraction, at an optimized force of 30,000 lbs for a residence time of 15 min to allow for effective separation. The green juice was stored in a freezer at −20 °C until use. Then, 100 g of deionized water was added to 50 g of the pressed miscanthus solid cake. The 2:1 mixtures were thoroughly mixed and chopped in a rotary knife mill Grindomix GM 200 (Retsch®, Verder Scientific Inc. Newtown, PA, USA) at a speed of 9000 RPM for 2 min. Subsequent juice separation was conducted using a centrifuge (Centra-GP8R Centrifuge, ThermoIEC, Champaign, IL, USA). The centrifugation was carried out at a rotational speed of 2600 RCF for 10 min at 25 °C. The solid cake was stored in sealable containers at −4 °C for further analysis and downstream processing. Figure 1 illustrates the procedure taken in this work, including mechanical separation, microalgae cultivation, thermochemical processing and bioethanol fermentations.

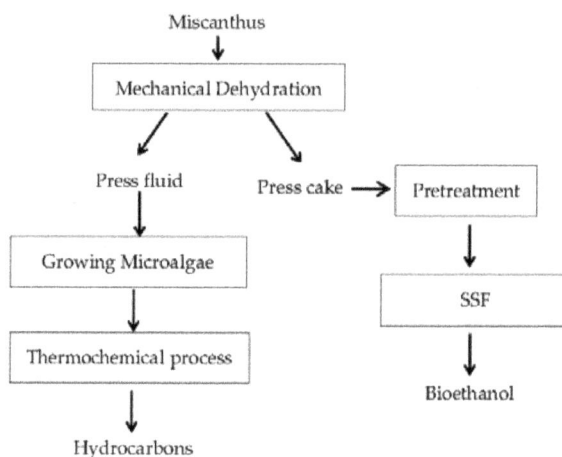

Figure 1. Flow chart of the integrated generation of bioethanol and hydrocarbons from uses of MxG' solid and liquid fractions. (SSF: Simultaneous saccharification and fermentation).

2.2. Biomass Analytical Procedures

The solid fraction and juice fraction of miscanthus after separation were analyzed for minerals (e.g., K, Mg, Ca, Cl, S, P), elemental composition (e.g., C, H, O, N), solids content, ash content, volatile content, and carbohydrates (cellulose, hemicellulose, lignin). Two stages of acid hydrolysis were performed for determining the structural carbohydrates and lignin content on the alfalfa samples according to NREL Ethanol Project Laboratory Analytical Procedure. In addition to chemical analyses, the mass flow of dry matter from the MxG into the juice and solid cake were calculated.

The elemental composition (C, H, O, N) of the solid cake samples was determined using a PE 2400 II CHNS/O analyzer (Perkin Elmer Japan Co., Ltd., Yokohama, Japan). The miscanthus samples were digested with HNO3/HCl in a microwave oven (200 °C, 2 MPa) and analyzed by inductively coupled plasma-optical emission spectroscopy (ICP-OES) (ARL 3560, Waltham, MA, USA) for minerals.

Compounds in MxG juice was determined using LC/MS. The concentration of monomeric sugars, including cellobiose, glucose, arabinose and xylose in all liquid fractions as well as the concentration of ethanol were all determined using a Dionex Ultimate 3000 (UHPLC) (Thermo Fisher Scientific, Bannockburn, IL, USA) equipped with a Shodex Sugar SH 1218 ion exclusion column and a Shodex RI-101 refractive index detector [17]. The samples were analyzed using a UPLC-QTOF-MS system (ACQUITY UPLC-SYNAPT MS, Waters Corp., Milford, MA, USA). An ACQUITY UPLC BEH C18 1.7 µM VanGuard pre-column and an ACQUITY UPLC BEH C18 1.7 µM analytical column were used for the analyses at column temperature of 40 °C and a flowrate of 0.4 (mL/min). The system was operated in both electrospray ionization (ESI) positive (mobile phases: 0.1% formic acid in water and 0.1% formic acid in acetonitrile) and negative modes (mobile phases 1 mM ammonium fluoride in water and acetonitrile) to provide comprehensive coverage. Metabolite annotation for UPLC-QTOF-MS analysis was performed by comparing mass spectra and retention time of each detected signal from the MxG samples to those of reference standards in an in-house library, which contained 500+ endogenous metabolites and plant metabolites.

The mobile phase was 0.01N H_2SO_4 at a flowrate of 1 mL/min. The temperature of the detector and column were maintained at 50 °C and 75 °C, respectively. All experiments and analyses were performed in duplicate.

2.3. Microalgae Cultivation

Chlorella vulgaris (UTEX 2714), which was obtained from the Culture Collection of Algae at the University of Texas at Austin, was used in this study. The MxG juice was diluted to DI water to prepare 1, 2, 5 and 10% (v/v) juice media. A 10 vol % algal inoculum was used for scaling-up the culture to a larger volume. Microalgae cultivation was carried out in 150 mL Wheaton glass bottles containing 100 mL algal culture (i.e., 90 mL of medium plus 10 mL of algal inoculum) at room temperature and 200 μmol m^{-2} s^{-1} continuous cool-white fluorescent light illumination. Agitation of the bottles was carried out manually once daily and the optical density (OD) of the microalgal culture was measured everyday using a Thermo Scientific GENESYS 20 Spectrophotometer (Waltham, MA, USA). A previously established correlation (Equation (1)) between the optical density of this particular *C. vulgaris* at 680 nm and the cell number was used for analysis [18].

$$\text{Cell number (cell/mL)} = 8 \times 10^6 \, \text{OD}_{680} + 425{,}897 \tag{1}$$

At the end of the microalgal cultivation, the microalgal broth was centrifuged at 2300× *g* and 20 °C for 15 min. Supernatants were separated, and the collected microalgal cells were dried at 60 °C for 48 h until the sample reached equilibrium moisture content.

2.4. Thermalchemical Conversion Analysis of Microalgae via Py-GC/MS and TGA

The pyrolysis-gas chromatography/mass spectrometry (Py-GC/MS) analysis of microalgal biomass was carried out in a Frontier EGA/PY-3030D multi-shot pyrolyzer (Fukushima, Japan), which was coupled with an Agilent 7890A gas chromatography/5975c mass spectrometer (GC/MS) with a DB-5MS capillary column (Santa Clara, CA, USA). For each experiment, the pyrolyzer was pre-heated to desired temperatures of 500 °C, following by dropping approximately 0.3 mg of sample into the pyrolysis part of the reactor. The sample was volatized immediately, giving off pyrolysis products as the vapor, which was injected directly into GC/MS. The compounds were identified by comparison with the mass spectral database of the National Institute of Standards and Technology (Gaithersburg, MD, USA).

Thermogravimetry analysis (TGA) was performed by using a SDT Q600 thermalgravimetric analyzer (TA Instruments, New Castle, DE, USA), in which the sample was heated to 600 °C at a heating rate of 5 °C/min in nitrogen.

2.5. Dilute Sulfuric Acid Pretreatment

A Dionex ASE 350 Accelerated Solvent Extractor (Dionex Corporation, Sunnyvale, CA, USA) was used to do the dilute sulfuric acid pretreatment. A 1% (v/v) sulfuric acid solution was prepared to pretreat the biomass, at a pretreatment temperature of 160 °C for 10 min. Approximately 30 g of blended wet miscanthus was placed into a tared 66 mL Dionex extraction cell containing a glass fiber filter. Then the appropriate number of 150 mL collection vials were weighed and placed onto the ASE system. The extractor passed 60 mL of dilute sulfuric acid solution into each cell containing the biomass. Then the cells were heated to the desired temperature (160 °C) at a heating rate of 25 °C/min, with the temperature maintained for 15 min. After this treatment, 40 mL of the dilute sulfuric acid solution was passed into the cells to rinse the biomass. The resulting extractive and the rinsing solution (total about 100 mL) were collected in the collection vials. The extraction cells were cooled down to 25 °C. The recovered biomass samples were stored in sealable containers and kept at −4 °C for further downstream processing and analyses such as compositional analysis and fermentation.

2.6. Fermentation of MxG Solid Fraction to Produce Ethanol

MxG solid cake was taken through simultaneous saccharification and fermentation (SSF) to produce ethanol using *S. cerevisiae* microorganisms. For SSF, a pretreated biomass loading of 5 g (wet basis) with a total working volume of 50 mL and a pH adjusted to 4.5 by the addition of 0.05 M citric buffer was used. Wheaton septum glass bottles (125 mL) were used as fermentation reaction vessels. A cocktail of enzymes including cellulose (Novozyme NS 50013) at a loading of 60 FPU/g glucan, hemicellulase (Novozyme NS 22002) at a loading of 2.5 FBG/g glucan and β-glucosidase (Novozyme NS 50010) at 4.5 CBU/g glucan was used for enzymatic hydrolysis. *Saccharomyces cerevisiae* (ATCC 24858) was then added to the reaction vessel to begin the SSF process.

The fermentation cultures were placed in a rotary shaker and incubated at 35 °C and 200 rpm and grown aerobically. Samples were taken at predetermined intervals (0, 3, 6, 12, 24, 48, 72, and 96 h) and collected by filtering through 0.45-μm nylon membranes for ethanol and sugars analysis by HPLC. *Saccharomyces cerevisiae* cultured with YM broth was also used for SSF to serve as control and allow for comparison. The ethanol yield was expressed as the percentage of the theoretical yield using the following formula:

$$\% \text{ Yield}_{ethanol} = \left[\frac{C_{ethanol,f} - C_{ethanol,i}}{0.568 f . C_{biomass}} \right] \times 100\% \qquad (2)$$

where $C_{ethanol,f}$ is the ethanol concentration at the end of the fermentation (g/L), $C_{ethanol,i}$ is the ethanol concentration at the beginning of the fermentation (g/L), $C_{biomass}$ is the dry biomass concentration at the beginning of the fermentation (g/L), f is the cellulose fraction of the dry biomass (g/g), and 0.568 is the conversion factor from cellulose to ethanol. All experiments were carried out in duplicates.

3. Results and Discussion

3.1. Characteristics of Raw Miscanthus, Miscanthus Cake and Juice

The mechanical press was largely effective in separating freshly harvested MxG into juice and cake with a percentage mass distribution of between 0.5 g/g of liquid and 0.5 g/g of solid [17]. The compositions of the separated MxG solid cake and juice are listed in Table 1. One of the most notable differences between the cake and juice was the significantly lower total solids content of the MxG juice (~0.1%). Differences also existed in the carbohydrates group (cellulose, hemicellulose, lignin) among the cake and juice samples. As expected, the carbohydrates of the cake are higher than those of the juice. This is mainly because the cellulosic fibers are less extractable due to the polysaccharide matrix structure in biomass [13]. The elemental compositions of the two fractions are comparable, with a higher carbon content of the solid cake. MxG has been noted to be a good candidate for bioethanol production due to its high carbon content. The juice fraction showed a higher amount of crude protein content than the solid cake, indicating proteins in the raw miscanthus were likely extracted in to the juice fraction.

According to the ICP analysis, the Calcium (Ca), Potassium (K), and Magnesium (Mg) concentrations of the MxG juice were much higher than the solid cake, while the Sodium (Na) concentrations of The MxG juice was significantly lower than the solid cake. Carbon, Nitrogen and Phosphorus are the three primary nutrients for algae cultivation. Other micronutrients also required include silica, calcium, magnesium, potassium, iron, manganese, sulfur, zinc, copper and cobalt [19].

The LC/MS further confirmed that the MxG juice contains trace amount of amino acids, organic acids, and metabolites, which are good feed supplements for microorganism growth (Table 2).

Table 1. Characteristics of MxG cake and juice.

Group/Specific	Miscanthus Cake	Miscanthus Juice
Total Solids, wt %	17.4	0.1
Ash, % dry matter	1.8	3.9
H_2O, wt %	82.6	99.9
Biomass Composition of Dry Matter, wt %		
Cellulose	47.5	39.2
Hemicellulose	36.6	19.7
Lignin	16.5	12.3
Protein Content	2.5	5
Elemental Composition (%)		
C	48.2	37.8
H	6.6	5.2
N	0.4	0.8
S	0.5	1.0
Mineral Composition(ppm)		
Al	21.3	39.3
B	52	95.3
Ca	412.3	1968.2
Cd	2.7	7.4
Cu	25.1	20.1
Fe	97.4	76.2
K	5506.6	93,083.3
Mg	1901.3	19,658.3
Mn	30.8	251.3
Na	1301.0	487.1
Ni	7.5	5.1
P	84.2	475.7
Pb	7.0	10.3
S	11.0	23.2
Zn	22.9	39.7
Si (g/mL)	2926.4	273.6
Mo (g/mL)	11.5	6.9

Table 2. LC/MS analysis of MxG juice.

Compound Name	Ret Time (min)	*m/z*
3-hydroxypropionic acid	0.6756	89.0
7-methylxanthine/3-methylxanthine	1.4121	167.1
Adenine	0.8119	136.1
aminobutyric acid	0.6186	102.1
apigenin/genistein	3.332	271.0
Arginine	0.7322	175.1
azelaic acid	2.9221	211.1
Biotin	2.2158	227.1
caffeic acid	2.2282	181.1
Citrulline	0.7176	159.1
Cytosine	0.6817	112.1
fumaric acid	0.5659	115.0
Genistin	2.392	433.1
gibberellic acid	2.7087	347.2
glucaric acid	0.6975	233.0

Table 2. *Cont.*

Compound Name	Ret Time (min)	*m/z*
gluconic acid	0.6487	195.0
glutamic acid	0.8481	148.1
Histidine	0.686	154.1
Hypotaurine	0.5547	110.0
Hypoxanthine	1.0334	137.0
Leucine	1.2646	132.1
L-tyrosine	1.1882	182.1
m-Couraric acid	2.6982	167.1
Methionine	0.9698	133.0
nicotinic acid	0.9439	124.0
Pantothenate	1.6058	220.1
Paraxanthine	1.9946	181.1
Phenylalanine	1.506	166.1
phosphoenolpyruvic acid	0.6185	167.0
Proline	0.7951	116.1
Riboflavin	2.1214	377.1
succinic acid	1.1895	119.0
Thymine	1.413	127.0
Tryptamine	2.0723	144.1
Tryptophan	1.794	188.1
Uracil	0.8922	113.0

3.2. Miscanthus Juice as a Nutrition Supplement for Microalgal Growth

Figure 2 shows the growth curves of *C. vulgaris* in the MxG juice media. As the control, DI-water alone cannot support the algal growth. The solid content of the MxG juice is 0.1 wt %, meaning that the actual solid contents in 1%, 2%, 5%, 10% and 15% juice media were 0.001%, 0.002%, 0.005%, 0.01% and 0.015%, respectively. It was observed that the higher concentration on MxG juice, the higher the microalgal growth rate. The highest cell density was observed in the 15% juice medium. A similar growth pattern was observed in the 1% and 5% juice medium. These results indicate that MxG juice was able to supple the microalgae with the necessary nutrients for its growth. Work done by Rahman et al. (2015), using cattail juice, showed that the cattail juice was similar to proteose medium, which was rich enough to support high-cell-density growth of numerous microalgae strains [18]. Juice extracts from plants therefore hold great potential as alternative growth media for microalgae and bacteria.

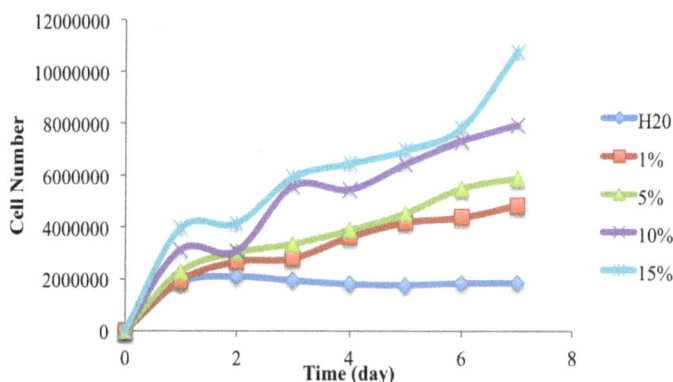

Figure 2. Growth curves of *C. vulgaris* in DI water with MxG juice at room temperature and 600 µmol m^{-2} s^{-1} light intensity.

3.3. Thermochemical Conversion of Microalgae Grown on MxG Juice

C. vulgaris grown on MxG juice contained approximately 64.6% protein, 8.9% carbohydrates and 12.3% lipids (Table 3). Its biochemical and elemental compositions were similar to the same species grown on other media [15].

Table 3. Elemental and composition analysis of microalgae (moisture free basis, % by weight).

Composition	C	H	N	S	Protein	Carbohydrates	Lipid	Volatile Solid	Ash
Chlorella	44.3	6.8	10.3	1.2	64.6	8.9	12.3	88.5	11.5

TGA study of *C. vulgaris* used a condition that was similar to slow pyrolysis. Figure 3 shows the Thermogravimetry (TG) and Differential thermogravimetry (DTG) curves of microalgae that exhibited its weight loss characteristics. The decomposition of this microalgal biomass can be divided into two phases. The first phase (T < 180 °C) was moisture removal. The decomposition phase occurred between 180 and 350 °C with a weight loss of 45 wt %. The results indicated that applying the slow pyrolysis process to *C. vulgaris* might result in a residue yield of 20.1 wt % of the starting material (i.e., the conversion ratio was 79.9 wt %). The possible reasons for the high residue yield are the ash content and the high protein content in feedstock.

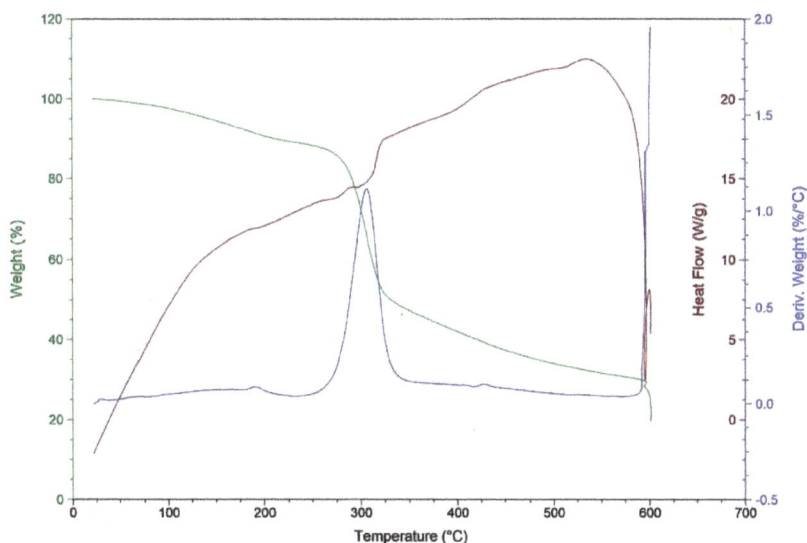

Figure 3. Thermogravimetry (TG) and DTG curves of microalgae.

Py-GC/MS analysis of microalgae *C. vulgaris* was performed at 500 °C, resulting in over 100 different chemicals (Figure 4). The top 20 pyrolytic products representing 48% (area) of all products are summarized in Table 4. Pyrolysis of microalgae could form a complex organic mixture. The chemicals in the mixture can be generally categorized as fatty acids (such as C16 hexadecenoic acid and C14 tetradecanoic acid), hydrocarbons (like toluene, butane, cresol and octadecane), nitrogenated compounds (such as pyridines, pyrazines, pyrroles, indoles and their derivatives), and oxygenated compounds (organic acids, ketones, furfurals, aldehydes and phenols). Fatty acids are degradation products of microalgal lipids, and C16 and C14 fatty acids are major components. Nitrogenated compounds are derived from microalgal proteins. Although a fair amount of hydrocarbons was produced via pyrolysis of microalgae, the products are still combined with hundreds of other chemicals. To use microalgal pyrolysis bio-oil as the transportation fuel, an upgrading process is still required [20].

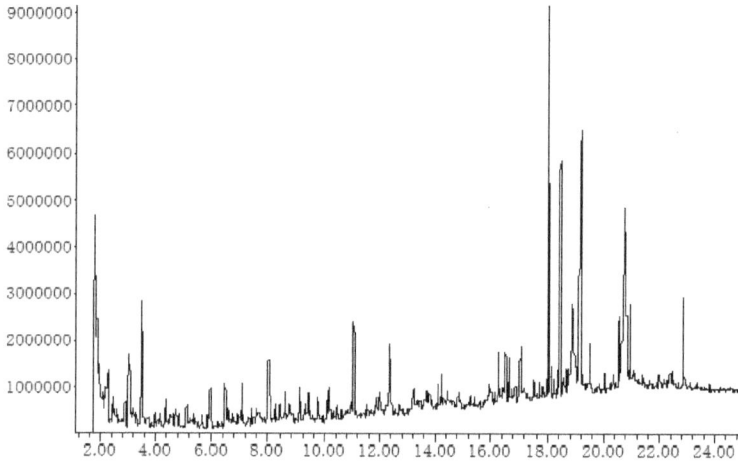

Figure 4. Gas chromatography (GC) profiles of pyrolytic products of microalgae.

Table 4. Pyrolysis-gas chromatography/mass spectrometry (Py-GC/MS) analysis of microalgae *C. vulgaris*.

Possible Chemical	Retention Time	Area %
Hexadecenoic acid	19.18	6.83
9-Octadecyne	18.03	6.48
9,12,15-Octadecatrienoic acid, (Z,Z,Z)-	20.77	6.14
Methyl 8,11,14-heptadecatrienoate	18.89	4.55
Toluene	3.47	2.93
Acetic acid	2.16	2.40
Methanethiol	1.84	2.22
Butane	1.91	2.14
N,N-Dimethylaminoethanol	3.00	2.10
p-Cresol	8.00	1.81
1H-Indole, 6-methyl-	12.37	1.67
Oleic Acid	20.81	1.45
Indole	11.07	1.13
Tetradecanoic acid	17.08	0.98
Tetradecanamide	20.97	0.96
Octadecane	22.86	0.96
1H-Pyrazole, 1,3,5-trimethyl-	5.95	0.93
Piperidine, 1-(cyanoacetyl)-	13.25	0.92
Butanal, 3-methyl-	2.44	0.86
Butanal	2.06	0.82

3.4. Effect of Pretreatment on the Cake of Miscanthus X Giganteus

The purpose of the pretreatment was to remove lignin and/or hemicellulose, to disrupt the crystalline structure of cellulose, and to increase the porosity of the material, making it more accessible to enzyme attack [21]. The compositions of miscanthus before and after dilute sulfuric acid pretreatment are given in Table 5.

Table 5. Composition of MxG before and after dilute sulfuric acid pretreatment.

Sample	Cellulose (%)	Hemicellulose (%)	Lignin (%)	Ash (%)
Untreated	49.34	32.75	15.25	1.23
Pretreated	71.8	1.27	26.83	0.52

Untreated MxG contained approximately 49% cellulose, 32.75% hemicellulose, and 15.25% lignin (Table 5). Miscanthus, with its cellulose content of approximately 40% and above, is considered useful for biofuel production [22]. The compositions of the raw miscanthus in the present study is consistent with previous findings reported so far [23].

After dilute sulfuric acid pretreatment, the composition of the biomass changed, with an increase in the cellulose and lignin contents and a decrease in the hemicellulose content the biomass samples. The hemicellulose of biomass decreased from about 32.75% in the untreated samples to approximately 1% in the pretreated samples, indicating almost complete hydrolysis of hemicellulose fraction of the biomass This removal of hemicellulose increases porosity and improves enzymatic digestibility. The complete removal of hemicellulose usually results in a maximum enzymatic digestibility [24], while the increase of the porosity of the sample can cause an increase in the lignin content of the sample. Studies of dilute acid pretreatment show the increase in the lignin content would not affect the ethanol yield and concentration [25–27].

3.5. Fermentation of MxG Cake for Ethanol Production

The MxG cake was treated with dilute acid first, and then used to produce bioethanol via a simultaneous saccharification and fermentation (SSF) process. Bioethanol was rapidly produced within the first 24 h, while the concentration of glucose reduced rapidly as well, with the measured xylose concentration remaining stable throughout the period (Figure 5). The production of bioethanol began to level off during the following 72 h, with a continued reduction in glucose concentration and a steadiness of the xylose concentration (Figure 5). The final ethanol yield was 88.4% of the theoretical value. This result suggests that sugars produced from the MxG cake can be efficiently fermented to ethanol.

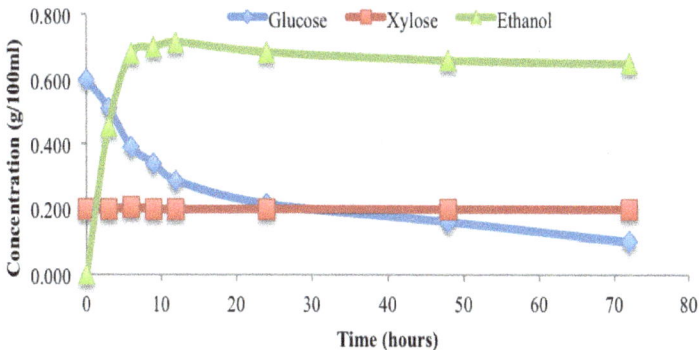

Figure 5. Ethanol, glucose and xylose profiles during the saccharification and fermentation (SSF) of the MxG cake.

4. Conclusions

Freshly harvested MxG was fractionated using mechanical press, whereby MxG was separated into a fiber-rich cake and a nutrient-rich juice. The mechanical pressing proved to be very efficient at reducing the solids mass transfer to the juice, resulting in a very low solid content of the juice.

The MxG juice was used to cultivate *C. vulgaris* in different media. Results showed the juice was highly nutritious and supported the growth of microalgal culture. Thus the juice could be a highly nutritious source for the microorganisms and bacteria. The uses of the MxG juice eliminate the extracted juice as a waste stream within a green biorefinery platform, making the concept more economical and environmentally friendly. Further, microalgae grown with the juice were studied for its thermochemical conversion behaviors via Py-GC/MS and TGA. Pyrolysis of this microalgae might result in approximately 79.9% conversion and a bio-oil containing over 100 chemicals.

The MxG cake was pretreated with dilute acid, and used as the feedstock for ethanol production through SSF using yeasts. Glucose from MxG solid cake can be efficiently fermented to ethanol with 88.4% of the theoretical yield. The dilute acid pretreatment was sufficient for pretreating MxG cake, resulting in very low hemicellulose content and high cellulose content of the pretreated cake.

Acknowledgments: This work is supported by the USDA National Institute of Food and Agriculture, (Evans-Allen) project (Grant No. NCX-272-5-13-130-1 and NC.X-303-5-17-130-1).

Author Contributions: Shuangning Xiu devised and drafted the manuscript; Bo Zhang performed the microalgae analysis and thermochemical conversion process. Nana Abayie Boakye-Boaten performed the experiments of pretreatment and bioethanol production and analyzed the data; Bo Zhang and Abolghasem Shahbazi revised the manuscript.

Conflicts of Interest: The authors declare no conflict of interest.

References

1. Cherubini, F. The biorefinery concept: Using biomass instead of oil for producing energy and chemicals. *Energy Convers. Manag.* **2010**, *51*, 1412–1421. [CrossRef]
2. Parajuli, R.; Dalgaard, T.; Jørgensen, U.; Adamsen, A.P.S.; Knudsen, M.T.; Birkved, M.; Gylling, M.; Schjørring, J.K. Biorefining in the prevailing energy and materials crisis: A review of sustainable pathways for biorefinery value chains and sustainability assessment methodologies. *Renew. Sustain. Energy Rev.* **2015**, *43*, 244–263.
3. McLaren, J.S. Crop biotechnology provides an opportunity to develop a sustainable future. *Trends Biotechnol.* **2005**, *23*, 339–342. [PubMed]
4. FitzPatrick, M.; Champagne, P.; Cunningham, M.F.; Whitney, R.A. A biorefinery processing perspective: Treatment of lignocellulosic materials for the production of value-added products. *Bioresour. Technol.* **2010**, *101*, 8915–8922. [CrossRef] [PubMed]
5. Kamm, B.; Kamm, M. Biorefineries—Multi product processes. In *White Biotechnology*; Ulber, R., Sell, D., Eds.; Springer: Berlin, Germany, 2007; pp. 175–204.
6. Xiu, S.; Shahbazi, A. Development of Green Biorefinery for Biomass Utilization: A Review. *Trends Renew. Energy* **2015**, *1*, 4–15. [CrossRef]
7. Khanna, M.; Dhungana, B.; Clifton-Brown, J. Costs of producing miscanthus and switchgrass for bioenergy in Illinois. *Biomass Bioenergy* **2008**, *32*, 482–493. [CrossRef]
8. Boakye-Boaten, N.A.; Xiu, S.; Shahbazi, A.; Wang, L.; Li, R.; Schimmel, K. Uses of miscanthus press juice within a green biorefinery platform. *Bioresour. Technol.* **2016**, *207*, 285–292. [CrossRef] [PubMed]
9. Wachendorf, M.; Richter, F.; Fricke, T.; Graß, R.; Neff, R. Utilization of semi-natural grassland through integrated generation of solid fuel and biogas from biomass. I. Effects of hydrothermal conditioning and mechanical dehydration on mass flows of organic and mineral plant compounds, and nutrient balances. *Grass Forage Sci.* **2009**, *64*, 132–143. [CrossRef]
10. Balat, M. Production of bioethanol from lignocellulosic materials via the biochemical pathway: A review. *Energy Convers. Manag.* **2011**, *52*, 858–875. [CrossRef]
11. Xiu, S.; Boakye-Boaten, N.A.; Shahbazi, A. Separate hydrolysis and fermentation of untreated and pretreated alfalfa cake to produce ethanol. In Proceedings of the 2013 National Conference on Advances in Environmental Science and Technology; Uzochukwu, G.A., Schimmel, K., Kabadi, V., Chang, S.-Y., Pinder, T., Ibrahim, S.A., Eds.; Springer: Cham, Switzerland, 2016; pp. 233–240.
12. Takara, D.; Khanal, S.K. Green processing of tropical banagrass into biofuel and biobased products: An innovative biorefinery approach. *Bioresour. Technol.* **2011**, *102*, 1587–1592. [PubMed]

13. Rahman, Q.M.; Wang, L.J.; Zhang, B.; Xiu, S.N.; Shahbazi, A. Green biorefinery of fresh cattail for microalgal culture and ethanol production. *Bioresour. Technol.* **2015**, *185*, 436–440. [CrossRef] [PubMed]

14. Andersen, M.; Kiel, P. Integrated utilisation of green biomass in the green biorefinery. *Ind. Crops Prod.* **2000**, *11*, 129–137. [CrossRef]

15. Arlabosse, P.; Blanc, M.; Kerfaï, S.; Fernandez, A. Production of green juice with an intensive thermo-mechanical fractionation process. Part I: Effects of processing conditions on the dewatering kinetics. *Chem. Eng. J.* **2011**, *168*, 586–592. [CrossRef]

16. Xiu, S.; Shahbazi, A.; Boakye-Boaten, N.A. Effects of Fractionation Methods on the Isolation of Fiber-Rich Cake from Alfalfa and Ethanol Production from the Cake. *BioResources* **2014**, *9*, 3407–3416. [CrossRef]

17. Boakye-Boaten, N.A.; Xiu, S.; Shahbazi, A.; Wang, L.; Li, R.; Mims, M.; Schimmel, K. Effects of fertilizer application and dry/wet processing of miscanthus x giganteus on bioethanol production. *Bioresour. Technol.* **2016**, *204*, 98–105. [CrossRef] [PubMed]

18. Hasan, R.; Zhang, B.; Wang, L.; Shahbazi, A. Bioremediation of swine wastewater and biofuel potential by using chlorella vulgaris, chlamydomonas reinhardtii, and chlamydomonas debaryana (corrected version). *J. Pet. Environ. Biotechnol.* **2014**, *5*, 175–180. [CrossRef]

19. Zhang, B.; Wang, L.; Hasan, R.; Shahbazi, A. Characterization of a native algae species chlamydomonas debaryana: Strain selection, bioremediation ability, and lipid characterization. *BioResources* **2014**, *9*, 6130–6140. [CrossRef]

20. Zhang, B.; Wang, L.; Li, R.; Rahman, Q.M.; Shahbazi, A. Catalytic conversion of chlamydomonas to hydrocarbons via the ethanol-assisted liquefaction and hydrotreating processes. *Energy Fuels* **2017**, *31*, 12223–12231. [CrossRef]

21. Oberoi, H.S.; Vadlani, P.V.; Saida, L.; Bansal, S.; Hughes, J.D. Ethanol production from banana peels using statistically optimized simultaneous saccharification and fermentation process. *Waste Manag.* **2011**, *31*, 1576–1584. [CrossRef] [PubMed]

22. Sørensen, A.; Teller, P.J.; Hilstrøm, T.; Ahring, B.K. Hydrolysis of miscanthus for bioethanol production using dilute acid presoaking combined with wet explosion pre-treatment and enzymatic treatment. *Bioresour. Technol.* **2008**, *99*, 6602–6607. [CrossRef] [PubMed]

23. Scordia, D.; Cosentino, S.L.; Jeffries, T.W. Effectiveness of dilute oxalic acid pretreatment of miscanthus × giganteus biomass for ethanol production. *Biomass Bioenergy* **2013**, *59*, 540–548. [CrossRef]

24. Ballesteros, I.; Ballesteros, M.; Manzanares, P.; Negro, M.J.; Oliva, J.M.; Sáez, F. Dilute sulfuric acid pretreatment of cardoon for ethanol production. *Biochem. Eng. J.* **2008**, *42*, 84–91. [CrossRef]

25. Zhu, L.; O'Dwyer, J.P.; Chang, V.S.; Granda, C.B.; Holtzapple, M.T. Structural features affecting biomass enzymatic digestibility. *Bioresour. Technol.* **2008**, *99*, 3817–3828. [CrossRef] [PubMed]

26. Boonsawang, P.; Subkaree, Y.; Srinorakutara, T. Ethanol production from palm pressed fiber by prehydrolysis prior to simultaneous saccharification and fermentation (SSF). *Biomass Bioenergy* **2012**, *40*, 127–132. [CrossRef]

27. Torr, K.M.; Love, K.T.; Simmons, B.A.; Hill, S.J. Structural features affecting the enzymatic digestibility of pine wood pretreated with ionic liquids. *Biotechnol. Bioeng.* **2016**, *113*, 540–549. [CrossRef] [PubMed]

fermentation

MDPI

Article

Integrated Process for Extraction of Wax as a Value-Added Co-Product and Improved Ethanol Production by Converting Both Starch and Cellulosic Components in Sorghum Grains

Nhuan P. Nghiem *, James P. O'Connor and Megan E. Hums

Eastern Regional Research Center, Agricultural Research Service, U.S. Department of Agriculture, Wyndmoor 19038, PA, USA; jpoconn@g.clemson.edu (J.P.O.); Megan.Hums@ARS.USDA.GOV (M.E.H.)
* Correspondence: john.nghiem@ars.usda.gov; Tel.: +1-215-233-6753

Received: 18 January 2018; Accepted: 11 February 2018; Published: 13 February 2018

Abstract: Grain sorghum is a potential feedstock for fuel ethanol production due to its high starch content, which is equivalent to that of corn, and has been successfully used in several commercial corn ethanol plants in the United States. Some sorghum grain varieties contain significant levels of surface wax, which may interact with enzymes and make them less efficient toward starch hydrolysis. On the other hand, wax can be recovered as a valuable co-product and as such may help improve the overall process economics. Sorghum grains also contain lignocellulosic materials in the hulls, which can be converted to additional ethanol. An integrated process was developed, consisting of the following steps: 1. Extraction of wax with boiling ethanol, which is the final product of the proposed process; 2. Pretreatment of the dewaxed grains with dilute sulfuric acid; 3. Mashing and fermenting of the pretreated grains to produce ethanol. During the fermentation, commercial cellulase was also added to release fermentable sugars from the hulls, which then were converted to additional ethanol. The advantages of the developed process were illustrated with the following results: (1) Wax extracted (determined by weight loss): ~0.3 wt % of total mass. (2) Final ethanol concentration at 25 wt % solid using raw grains: 86.1 g/L. (3) Final ethanol concentration at 25 wt % solid using dewaxed grains: 106.2 g/L (23.3% improvement). (4) Final ethanol concentration at 25 wt % solid using dewaxed and acid-treated grains (1 wt % H_2SO_4) plus cellulase (CTec2): 117.8 g/L (36.8% improvement).

Keywords: sorghum grains; fermentation process; fuel ethanol; sorghum wax; value-added co-products

---

1. Introduction

Ethanol has attracted attention worldwide as a clean and renewable liquid fuel. The recent low oil prices encouraged record gasoline consumption, which translated to increased demand for ethanol for use in E10 (10% ethanol, 90% gasoline) as well as higher blends such as E15 and E85. The United States currently is the largest producer of fuel ethanol in the world. Ethanol production in the United States reached a record of 57.73 billion liters (15.25 billion gallons) in 2016 [1]. More than 90% of all the ethanol produced in the United States comes from corn. Since corn prices tend to fluctuate [2], other starch-based feedstocks have been considered. Among these, grain sorghum has attracted strong interest because of high starch contents, which are equivalent to those of corn; low requirements for water and fertilizer; and high heat and drought tolerance. These characteristics allow sorghum to be grown in dry climates and regions where corn cannot thrive [3]. Currently, sorghum is used in at least nine commercial ethanol plants in the United States, mostly as an adjunct feedstock to corn [1].

The majority of commercial corn ethanol plants in the United States employ the dry-grind process. The economics of this process greatly depend on the revenue obtained by selling of the main co-product known as distillers dried grains with solubles (DDGS). Attempts have been made to develop additional co-products such as corn oil and corn fiber. Distillers corn oil is now an established co-product of almost all corn ethanol plants in the United States. In the case of sorghum, wax has been considered as a potential high-value co-product of ethanol production [4,5]. Sorghum wax has been found to possess several characteristics very similar to carnauba wax [4,6], which is an industrial wax with a wide range of applications. Carnauba wax is derived from the leaves of the *Copernicia prunifera* tree, which is found exclusively in Brazil [7]. Because of similar physical properties, sorghum wax has been suggested as a potential replacement for carnauba wax [4]. The bulk price of carnauba wax is listed at $6.65–$7/kg [8]. The global market of carnauba wax in 2015 was estimated at $246 million [9] and is expected to increase to $335 million in 2024 [10]. If sorghum wax can be recovered and used as a replacement for carnauba wax, it will open a relatively large market for a new co-product and potentially improve the economics of ethanol production using sorghum as feedstock. Cuticular waxes extracted from stalks of the sorghum plant have been found to inhibit acetone–butanol–ethanol fermentation [11]. The effect of grain sorghum wax on fermentation processes, however, has not been reported in the literature. It is possible that sorghum grain wax also has inhibitory effects toward ethanol fermentation. If this is the case, extraction of wax from the grains prior to fermentation will serve two purposes—development of a new high-value co-product and improvement of ethanol yield.

Similar to corn and other cereal grains, sorghum grain consists of an outer seed cover or pericarp, which encloses the embryo and the starch-rich endosperm [12]. The sorghum pericarp is lignocellulosic material, which can serve as feedstock for additional ethanol production. Prior to enzymatic hydrolysis for fermentable sugar production, lignocellulosic feedstocks have to be pretreated to increase the sugar yields. The pretreatment process employs various reagents and chemicals, which include high-pressure steam, acids, bases, organic solvents, and oxidizing agents such as ozone and hydrogen peroxide. The advantages and disadvantages of these pretreatment methods have been reviewed in detail [13].

In this paper, we report the development of an integrated process for extraction of sorghum grain wax as a high-value-added co-product and production of ethanol from the dewaxed sorghum using both starch and lignocellulosic components to improve ethanol yields.

2. Materials and Methods

2.1. Materials

2.1.1. Sorghum Grains

The sorghum grains were obtained from various sources and are listed in Table 1. All grains were free of pesticides. Upon receipt, the grains were stored in closed plastic containers and kept in the laboratory at ambient temperature until use.

Table 1. Sources and descriptions of sorghum grains.

Source	Description
Bob's Red Natural Foods (BRM) (Milwaukee, OR, USA)	Commercial product; unknown variety
United Sorghum Checkoff (USC) (Lubbock, TX, USA)	Blend of two varieties, Sorghum Partner SP6929 and Terral RV9782; the majority is SP6929 but the exact proportion is unknown
DuPont Pioneer (DPP) (Johnston, IA, USA)	Pioneer 83P56
Chromatin (Chicago, IL, USA)	SP7715

2.1.2. Enzymes and Chemicals

Spezyme® XTRA (thermostable α-amylase, activity 14,000 U/g), Fermenzyme® L-400 (glucoamylase/protease mix, activity 350 glucoamylase U/g) and Accellerase® 1500 (cellulase/β-glucosidase mix, endoglucanase activity 2200–2800 carboxymethylcellulase U/g; β-glucosidase activity 525–775 pNPG U/g where 1 pNPG unit liberated 1 μmol nitrophenol from *p*-nitrophenyl-β-D-glucopyranoside per min at 50 °C and pH 4.8) were provided by DuPont Industrial Biosciences (Palo Alto, CA, USA). Cellic® CTec2 (cellulase/β-glucosidase) was provided by Novozymes (Franklinton, NC, USA). Specific activities of CTec2 were not publicly disclosed by the manufacturer. All enzymes were kept refrigerated at 4 °C.

Active Dry Ethanol Red was provided by Lesaffre Yeast Corporation (Milwaukee, WI, USA). The dry yeast powder was kept refrigerated at 4 °C.

All chemicals were of reagent grade and purchased from various suppliers.

2.2. Methods

2.2.1. Wax Extraction

Sorghum grains were screened to remove small debris and broken pieces. Only intact and undamaged grains were used in the experiments. The moisture contents of the grains were determined prior to their use in the experiments. For wax extraction, 200.0 g (fresh weight) sorghum was placed in a 2 L flask containing 320 mL absolute ethanol. The mixture was vigorously stirred with a magnetic stir bar and heated on a heating plate. A glass condenser using laboratory cold water as a cooling medium was mounted on the top of the flask to provide total reflux of ethanol. After the ethanol started to boil, stirring and heating of the sorghum-ethanol mixture was continued for 30 min. During this time, the clear ethanol gradually became cloudy due to the presence of a milky white substance, which was later determined to be wax. The flask then was removed from the heating plate and allowed to cool to ambient temperature. The mixture was poured into a Buchner funnel to separate the wax-containing ethanol from the grains. The wax-containing ethanol was transferred to a beaker and about half of the ethanol was allowed to evaporate at ambient temperature. Samples were taken from the partially concentrated wax-ethanol mixture for use in the wax analysis. The recovered grains were washed with about 100 mL deionized (DI) water and dried in a 55 °C oven until constant weights were reached to determine the weight loss due to wax extraction.

2.2.2. Dilute H_2SO_4 Treatment

Bob's Red Mill (BRM) sorghum was selected for investigation of dilute H_2SO_4 treatment as a possible method to improve ethanol production. A batch of 200 g BRM sorghum was dewaxed as described previously. After the wax-containing ethanol was drained completely, the grains were transferred to a 500 mL glass media bottle and 180 mL of a dilute H_2SO_4 solution was added. Two acid concentrations—1 wt % and 2 wt %—were used in the study. The amount of acid solution used was sufficient to cover all the grains and still left about 100 mL of free liquid above them. The bottle was closed with the plastic cap, which was screwed into place then slightly loosened by about one tenth of a turn. The bottle was placed in an autoclave set at 121 °C for one hour. The bottle with its contents was weighed before and after autoclaving to determine the amount of water lost due to evaporation. The acid-treated grains were used directly for ethanol production without additional processing.

2.2.3. Ethanol Fermentation of Raw and Dewaxed Sorghums

Mashing

The raw and dewaxed sorghums were ground in a Krups model 203 coffee grinder (Solingen, Germany). The moisture contents of the sorghum meals were determined to calculate the amounts needed for preparation of the fermentation mash. In a 1 L stainless steel beaker, 125 g (dry weight) sorghum meal was mixed with DI water needed to make a mash with a total weight of 500 g. The total solid content of the mash was therefore 25 wt % on dry basis. The pH of the mash was adjusted to 5.6 using 5N H_2SO_4, and 34.1 µL of Spezyme Extra (0.3 kg enzyme/MT dry solids) was added. The beaker was placed in a hot oil bath for 2 h at 60 °C then 1 h at 90 °C to complete the starch liquefaction. Mixing of the mash was provided by a mechanical agitator. DI water was added throughout liquefaction to moderate viscosity in compensation for evaporation. After liquefaction, the beaker was cooled to 40 °C and weighed to determine the amount of water lost due to evaporation. DI water then was added to bring the total weight back to 500 g. The pH of the mash was adjusted to 4.0 using 5N H_2SO_4. After pH adjustment, 0.2 g urea and 73.9 µL Fermenzyme L-400 (0.65 kg enzyme/MT dry solids) were added. In the case of the BRM sorghum, two additional sets of experiments were performed. In these experiments, the enzyme dosages were increased to 2× and 5× of the aforementioned dosages.

Simultaneous Saccharification and Fermentation

Following urea and enzyme additions, the mash was stirred thoroughly to ensure complete dissolution of urea and uniform distribution of the enzyme. The mash then was split equally into six 250 mL flasks, each containing 50 g of mash. The active dry yeast was rehydrated by addition of 2.5 g to 50 mL DI water and stirred for 30 min. Each flask was inoculated with 0.25 mL of the yeast slurry. The flasks were capped with rubber stoppers which were pierced with 18 gauge hypodermic needles to allow for the release of CO_2. The flasks were incubated in an orbital shaking incubator maintained at 32 °C and 200 rpm. Simultaneous saccharification and fermentation (SSF) was performed for 72 h. During this time, the flasks were periodically weighed to determine the weight loss due to CO_2 production. The weight loss data were used to confirm that all fermentations were complete at 72 h. Final samples were taken from each flask and centrifuged on a microcentrifuge. The supernatants were filtered through a 0.2 micron filter into closed vials and stored in a freezer for analysis of residual sugars, ethanol, and other metabolite products.

Viscosity Reduction

In a separate set of experiments performed to study the effect of viscosity, at the end of the liquefaction, the beaker was cooled to 55 °C and 3.75 mL Accellerase 1500 was added (i.e., 0.03 mL/g solid). The mash was maintained at 55 °C and continuously stirred for 1 h. The rest of the mashing and fermentation procedure then followed.

2.2.4. Ethanol Fermentation of Dewaxed and H_2SO_4-Treated Sorghums

The dewaxed and H_2SO_4-treated BRM sorghum was used directly for mash preparation and subsequent SSF without washing and grinding. After dilute H_2SO_4 treatment, the grains were transferred from the media bottle to the stainless steel beaker for mashing. It was observed that a significant portion of the grains was liquefied into a thick slurry during pretreatment. Therefore, the entire contents of the bottle was used during mashing in order to maintain an accurate solid loading. The water content of the treated grains was calculated from the initial total mass and the water loss due to evaporation, which was determined by weighing the bottle before and after the acid treatment as described previously. DI water was added to the beaker until a 25% solid loading of grains was achieved. The mashing and SSF procedures were the same as described previously for the raw and dewaxed sorghums. To generate glucose from the cellulosic component of the grains for additional ethanol production, CTec2 was added at 0.03 mL/g solid.

2.3. Analytical Methods.

2.3.1. Moisture Determination

The moisture contents of whole sorghum grains were determined by placing 10 ± 2 g of material in preweighed aluminum weight boats. The boats were then dried in a 55 °C oven overnight and reweighed to determine the moisture loss. The moisture content was determined by calculating the amount of moisture loss as a percentage of the initial weight. The moisture contents of sorghum meals were determined by drying 2–3 g of material in an Ohaus MB45 moisture balance (Parsippany, NJ, USA). All measurements were performed in triplicate.

2.3.2. Wax Characterization

The partially concentrated wax-containing ethanol extracts were dried under nitrogen and dissolved in chloroform to make 5 mg/mL solution. Not all of the material was soluble in chloroform so the insoluble particles were removed using a syringe filter. The chloroform soluble fraction was dried under nitrogen and redissolved in chloroform to make a 5 mg/mL solution. The samples were analyzed using reverse-phase High Performance Liquid Chromatography (HPLC) with an Evaporative Light Scattering Detector (ELSD). HPLC analysis was performed using a Prontosil 200-3-C30 column (3.0 μm, 2.0 mm × 150 mm; Leonberg, Germany) on an Agilent 1260 series HPLC with an Agilent 1290 Infinity II series ELSD (Santa Clara, CA, USA). The method used was similar to one previously reported by Harron et al. [6] but was shortened from 95 min to 30 min by changing solvent gradients. Solvent A contained 99.9% methanol and 0.1% formic acid (v/v) while solvent B contained 99.9% chloroform and 0.1% formic acid (v/v). The flow rate was 0.200 mL/min with a column temperature of 50 °C. The mobile phase was initially 80:20 (solvent A/solvent B, v/v) and increased linearly to 20:80 over 10 min. The mobile phase was held at 20:80 from 10 min to 20 min and then was decreased linearly back to 80:20 by 21 min. The mobile phase was held at 80:20 from 21 min to 30 min. The ELSD was operated with an evaporator temperature of 80 °C, nebulizer temperature of 50 °C, and gas flow rate of 1.6 standard liters per minute (SLM).

2.3.3. Starch Determination

Starch contents of the raw and dewaxed sorghum grains were performed using the modified Megazyme assay [14]. The assay was based on the hydrolysis of starch with thermostable α-amylase and glucoamylase to produce glucose, which subsequently was determined and used for calculation of the starch content in the sample. The only modification was that the glucose produced in the present study was determined by an YSI glucose analyzer (YSI Incorporated, Yellow Springs, OH, USA) instead of a wet chemistry method as described in the Megazyme assay. All analyses were performed in triplicate.

2.3.4. Compositional Analysis of BRM Sorghum Fiber

The composition of the BRM sorghum fiber was determined according to the National Renewable Energy Laboratory (NREL) procedure [15]. To avoid interference by glucose from the starch in the grains, starch in the samples was removed prior to the compositional analysis. About 5 g BRM sorghum meal was placed in a 50 mL centrifuge tube and 40 mL 50 mM citric acid buffer at pH 4.8 was added followed by 0.5 mL Spezyme Extra. The tube was tightly capped, thoroughly mixed, and placed in an oven at 90 °C for 3 h. Every 30 min the tube was removed, thoroughly mixed, and replaced in the oven. At the end of the incubation period, the tube was cooled to ambient temperature, the pH of the slurry was adjusted to 4.5 with 5N H_2SO_4, and 0.5 mL Fermenzyme L-400 was added. The tube was again thoroughly mixed and placed in an incubator at 55 °C overnight (about 16 h). The tube then was centrifuged at 2500 rpm for 20 min. The supernatant was discarded and the pellet was washed three times, each time with about 50 mL DI water. The washed pellet was dried in the 55 °C oven and used for the compositional analysis. The analysis was performed in triplicate.

2.3.5. Analysis of Fermentation Samples

Residual glucose, ethanol, and other fermentation minor products were determined by HPLC. The system was an Agilent Technologies (Santa Clara, CA, USA) series 1200 equipped with a refractive index (RI) detector. The column was an Aminex® HPX-87H (Bio-Rad Laboratories, Hercules, CA, USA) operated at 60 °C. The solvent was 0.5 wt % H_2SO_4 pumped at a flow rate of 0.6 mL/min. Each sample was injected twice and the average results are reported.

3. Results and Discussion

3.1. Wax Extraction and Characterization

Wax could be extracted with a wide range of solvents such as hexane, benzene, chloroform, light petroleum ether, or acetone [5]. In the present study, ethanol was selected as the solvent for wax extraction because it is the final product in an ethanol plant and is readily available. The use of ethanol will therefore eliminate the solvent cost and reduce the total operating cost. In addition, the spent ethanol can easily be recovered in the distillation unit of the plant, either as a single stream or in combination with the fermentation-derived ethanol. The results of the wax extraction are summarized in Table 2.

Table 2. Wax contents of four sorghum varieties determined by weight losses during ethanol extraction.

Sorghum Type	Initial Dry Weight (g)	Wax Extracted (% of Initial Weight)
BRM	176.64	0.29
USC	171.86	0.19
DPP	174.26	0.17
Chromatin	174.51	0.29

The relative quantities of wax shown in Table 2 are calculated from the weight losses observed during the extraction process. The results indicate that the quantities of extracted waxes are approximately 0.2–0.3% of the initial mass of the sorghum grains, which is similar to those previously reported [5].

The chromatograms of the extracted waxes are superimposed and shown in Figure 1. In this method, oils such as triacylglycerol (TAG) eluted between 6–11 min while waxes eluted between 11–16 min. The chromatograms demonstrate the similarity of the four extracted waxes. The waxes in this chromatogram are similar to those identified by Harron et al. [6] and are primarily composed of C28-C30 fatty alcohols and aldehydes. The different peaks refer to mixtures of various chain length and saturation of the wax compounds, but for this analysis were not specifically analyzed by mass spectroscopy. The extracted waxes also show high purity and are practically free of oils (indicated by the extremely small peaks eluting between 6–11 min).

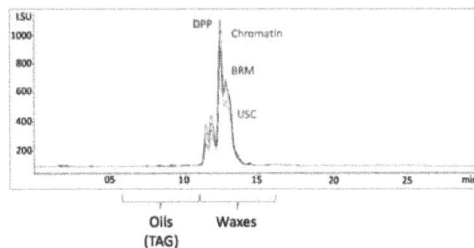

Figure 1. Chromatograms of the waxes extracted from the four sorghum grains: (1) Bob's Red Mill (BRM, blue), (2) United Sorghum Checkoff (USC, pink), (3) DuPont Pioneer (DPP, red), and (4) Chromatin (green) Oils (primarily triacylglycerol, TAG) eluted between 6-11 min and waxes eluted between 11–16 min.

3.2. Starch Contents

The results of starch content determination are shown in Table 3.

Table 3. Starch contents of raw and dewaxed sorghums.

Treatment	Starch (wt % Dry Basis)
BRM Raw	59.7 ± 3.0
BRM Dewaxed	67.4 ± 2.1
USC Raw	58.4 ± 7.8
USC Dewaxed	70.2 ± 3.5
DPP Raw	59.1 ± 1.8
DPP Dewaxed	65.3 ± 0.6
Chromatin Raw	58.7 ± 3.8
Chromatin Dewaxed	64.7 ± 1.6

From the results in Table 3 it can be seen that dewaxing resulted in significant increases in the measured starch contents, ranging from 10.1% for Chromatin to 20.0% for USC sorghum. The lower starch contents measured for the raw samples could probably be attributed to the inhibition of the starch hydrolytic enzymes used in the assay by the waxes. This is rather surprising since in the assay relatively low solid loading (about 3%) and very high enzyme dosages (about 3000 units/g solid of both enzymes) were used. The amounts of wax present in the assay mixture, therefore, seemed to be insufficient to cause the observable negative effect on the enzyme activities. Nevertheless, no other reasonable explanation could be provided. As discussed previously, there has been only one report on the negative effect of wax derived from sorghum stalk on a fermentation process [11]. There has been no report on the inhibition of either an enzymatic or fermentation process by grain sorghum wax. The result obtained in the present study is the first reported observation of a negative effect of grain sorghum waxes on an enzymatic process. The mechanism of this inhibition was not clear. Among the four sorghum grains tested, the DPP had a reddish brown color, which indicated the possible presence of tannin. To confirm its presence, whole and ground DPP grains were subjected to tannin extraction using a solution of 10 mM ascorbic acid in methanol at a solid/liquid ratio of 1:3 [16]. The extraction was performed at 55 °C for 16 h. The resultant light reddish brown color of the solvent indicated the presence of solubilized tannin. The extracted tannin was not quantified. Since ethanol extraction would remove only about 5% of tannin [16], most of the tannin was expected to remain in the DPP grains. The adsorption of proteins on tannin, which would severely limit their availability, has previously been reported in the literature [17]. The small increase of 10.5% of the measured starch content obtained for the dewaxed DPP over the raw grains indicate that the tannin in the DPP sorghum did not have significant effect on the enzymatic starch hydrolysis.

3.3. SSF of Raw and Dewaxed Sorghum Grains

The results of ethanol fermentation of the raw and dewaxed sorghum grains are shown in Table 4.

With the exception of Chromatin, dewaxing of sorghum grains resulted in increased ethanol production. These improvements can probably be linked to the higher efficiency of starch hydrolysis, which was the result of the wax removal. The highest increase of ethanol production, of 23.3%, was obtained with the BRM sorghum. For the USC sorghum, the increase of ethanol production was lower, at 5.6%. The smallest increase, of 2.8%, was observed for the DPP sorghum. As discussed in the previous section, the tannin in the DPP sorghum did not seem to have a negative effect on the starch hydrolysis. Whether the tannin had any negative effect on the fermentation is not known. The reason for no improvement of ethanol production in the case of the Chromatin sorghum is not clear. In all cases, addition of Accellerase 1500 slightly improved ethanol production. It was qualitatively observed that the addition of this cellulase enzyme formulation considerably reduced the viscosity of the mash, thus improving mixing, which would in turn improve both starch hydrolysis and fermentation.

Table 4. Final ethanol concentrations in the simultaneous saccharification and fermentation (SSF) of raw and dewaxed sorghum grains.

Feedstock	Final Ethanol (g/L)	Yield (% Theoretical)
BRM Raw	86.1 ± 2.0	71.1
BRM Dewaxed	106.2 ± 0.6	90.2
BRM Dewaxed with Accellerase	107.8 ± 1.9	91.8
USC Raw	102.2 ± 0.9	83.9
USC Dewaxed	107.9 ± 0.5	89.3
USC Dewaxed with Accellerase	108.2 ± 0.9	89.6
DPP Raw	104.1 ± 1.0	85.1
DPP Dewaxed	107.0 ± 3.8	87.8
DPP Dewaxed with Accellerase	109.6 ± 1.0	90.3
Chromatin Raw	98.7 ± 1.5	80.8
Chromatin Dewaxed	98.0 ± 5.4	80.2
Chromatin Dewaxed with Accellerase	99.7 ± 1.0	81.7

3.4. SSF of Dewaxed and H_2SO_4-Treated BRM Sorghum

Since the BRM sorghum gave the highest increase of ethanol production upon dewaxing, it was selected for investigation of potential further ethanol yield improvement by dilute H_2SO_4 treatment. To determine the potential additional ethanol yield, the compositions of the fiber obtained after destarching of the ground sorghum were determined. The results of the compositional analysis are shown in Table 5.

Table 5. Composition of the BRM sorghum fiber.

Component (wt % of Total Mass, Dry Basis)					
Glucan	Xylan	Arabinan	AI Lignin	AS Lignin	Ash
56.9 ± 1.6	4.9 ± 0.1	3.6 ± 0.0	10.7 ± 0.8	2.1 ± 0.0	0.1 ± 0.0

Notes: AI: Acid insoluble; AS: Acid soluble.

The high content of glucan in the fiber is favorable for ethanol production since upon hydrolysis it will result in high concentrations of glucose, which is the sugar most effectively fermented by the currently used commercial fuel-ethanol-producing *Saccharomyces cerevisiae* strains.

As discussed previously, the dewaxed and H_2SO_4-treated sorghum grains were used directly for mash preparation without washing and grinding. It was observed in the mashing process that the grains were sufficiently softened by the dilute acid treatment and slowly disintegrated when the mash was heated and agitated. The use of dilute acid for treatment of the sorghum grains offered an additional advantage. Prior to the start of the mashing process, the pH of the mash normally had to be adjusted to 4.5 with 5N H_2SO_4. The initial pH of the dewaxed and H_2SO_4-treated sorghum mash was found to be very close to the required pH. Therefore, only minimal pH adjustment was needed.

The final ethanol concentrations and the calculated yields obtained with the dewaxed and H_2SO_4-treated sorghums are shown in Table 6. The results obtained with the raw sorghum and dewaxed sorghum without H_2SO_4 treatment are also included for comparison. In all cases, the yields are calculated based on the total glucose available from both starch and cellulose. The results clearly demonstrate the improvements of ethanol production by treatment of the dewaxed sorghum with dilute H_2SO_4 and addition of the cellulase enzyme formulation CTec2, which resulted in the availability of more glucose for fermentation. Between the two acid concentrations used, 1 wt % H_2SO_4 gave slightly better ethanol yield than 2 wt % H_2SO_4. The higher acid concentration probably resulted in the formation of inhibitory compounds, which could have negative effects on the fermentation process. Compared with the raw sorghum, dewaxing and treatment with 1 wt % H_2SO_4 resulted in a 36.8% increase of ethanol yield.

Table 6. Final ethanol concentrations obtained in SSF of raw, dewaxed, and dewaxed plus dilute H_2SO_4-treated and cellulase-treated BRM sorghums.

Experiment	Final Ethanol (g/L)	Yield (% Theoretical) *
Bob's Red Mill raw	86.1 ± 2.0	62.7
Bob's Red Mill dewaxed	106.2 ± 0.6	77.4
Bob's Red Mill dewaxed and treated with 1 wt % sulfuric acid with CTec2 addition in SSF	117.8 ± 0.1	85.7
Bob's Red Mill dewaxed and treated with 2 wt % sulfuric acid with CTec2 addition in SSF	112.5 ± 4.5	82.0

* Based on total glucose available from starch and cellulose. A sample calculation of ethanol yield is shown in Appendix A.1.

3.5. The Proposed Integrated Process

Based on the results presented and discussed in the previous sections, an integrated process is proposed for ethanol production using grain sorghum as feedstock. In this process, glucose is obtained from both starch and fiber. The proposed process is shown in Figure 2.

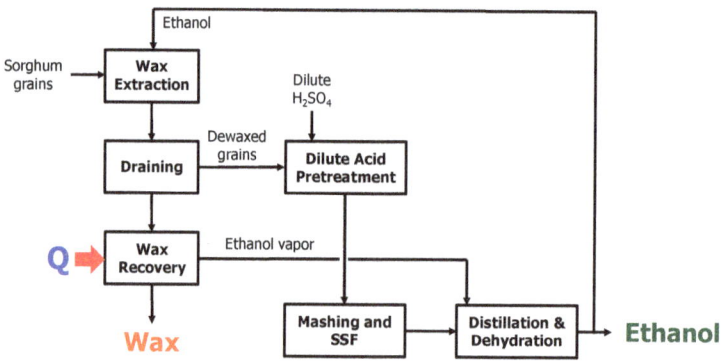

Figure 2. The integrated process for ethanol production from sorghum starch and fiber.

First, wax is extracted from the grains using ethanol, which is readily available in the plant. After a simple solid/liquid separation step—for example, by draining of the liquid—the extracted wax is recovered by evaporation of ethanol. The ethanol stream is brought to the distillation unit, where it can be fed to the distillation columns as a separate stream, or combined with other streams from the fermentation process, for ethanol recovery. The recovered ethanol is recycled and returned to the front for use in the next wax extraction cycle. The dewaxed grains are subjected to dilute H_2SO_4 treatment, then are used directly in mash preparation and subsequent SSF without washing and grinding. Omission of grain washing will result in significant savings of water consumption, whereas omission of grinding will result in significant savings of energy. The proposed process is simple and does not require expensive equipment. In addition, the key processing steps can be operated at moderate temperatures and pressures. The proposed process, therefore, can be added to an existing sorghum ethanol plant as a "bolt-on" process.

To realize the potential benefits, a simple economic analysis was performed for an ethanol plant producing 50 million gallons per year using the experimental data obtained with the BRM sorghum; the results are summarized in Table 7.

Table 7. Potential economic benefits of the integrated process for ethanol production using BRM sorghum grain as feedstock.

	Base Case	Dewaxed Only	Dewaxed/H$_2$SO$_4$ Treated
Ethanol yield (gal/bu)	2.04	2.59	2.92
Sorghum feedstock needed (million bu)	24.5	19.3	17.1
Total wax co-product (MT)	0	1252	1110
Wax co-product value (million $)	0	7.5	6.6
Sorghum feedstock savings (million $)	0	16.1	22.9

Notes: MT: metric ton; gal/bu: gallons per bushel; The value of the wax co-product is calculated using a bulk selling price of $6/kg; The feedstock savings are calculated using a unit cost of $3.10/bu. A sample calculation of potential economic benefits is shown in Appendix A.2.

The results in Table 7 indicate that if only wax extraction is performed, ethanol yield will increase from 2.04 to 2.59 gallons per bushel (gal/bu), which will translate into a saving of $16.1 million per year on reduced feedstock requirement. In addition, the wax co-product will add $7.5 million per year to the total revenue. If dilute sulfuric acid pretreatment also is performed on the dewaxed sorghum, the value of the wax co-product will decrease to $6.6 million per year but the ethanol yield will increase to 2.92 gal/bu and the saving on feedstock cost will increase to $22.9 million per year.

4. Conclusions

It has been demonstrated that wax could be extracted from four different commercial grain sorghum products by a simple process using ethanol as the solvent. The dewaxed sorghum grains gave higher ethanol yields when used as feedstock in an SSF process. Further improvements on ethanol yield were obtained when the dewaxed BRM sorghum grains were also treated with dilute solutions of sulfuric acid and a commercial cellulase was added during the fermentation to produce additional glucose from the fiber. Based on the experimental data, an integrated process for extraction of sorghum wax as a value-added co-product and production of ethanol from both starch and fiber was proposed. Using the data obtained for the BRM sorghum grains, the integrated process was shown to have significant economic benefits over the base process using raw sorghum as feedstock. The data, however, were obtained for proof-of-concept purpose only. More rigorous process optimization is needed and the benefits of dilute acid treatment must be proven with other sorghum types before the proposed integrated process is considered for commercial implementation.

Acknowledgments: The authors would like to thank DuPont Industrial Biosciences and Novozymes for providing the enzymes used in this study. The invaluable technical assistance of Gerard Senske, Matthew Toht, Michael Powell, and Katherine Norvell is sincerely appreciated.

Author Contributions: Nhuan P. Nghiem and James P. O'Connor contributed to the design and execution of the experiments, and data analysis. Megan E. Hums performed the wax characterization. All three authors contributed to the preparation and revision of the manuscript.

Conflicts of Interest: The authors declare no conflict of interest.

Appendix A.

Appendix A.1. Calculations of Ethanol Yields

Basis: 1000 g of 25 wt % raw BRM sorghum mash.

Total starch available: 250 g × 0.701 = 175.25 g.

Water consumed by hydrolysis of starch: 175.25 g × 0.11 = 19.28 g or 19.28 mL since the density of water is 1.

Let V_F be the final liquid volume in mL. Since the final ethanol concentration was 86.1 g/L, the additional volume (in mL) contributed by the ethanol produced was

0.0861 g/mL × V_F (mL) ÷ 0.789 g/mL = 0.109 V_F

Mass balance will give

V_F = 750 mL − 19.28 mL + 0.109 V_F

Therefore, V_F = 820.2 mL.

Total ethanol production: 86.1 g/L × 0.820 L = 70.62 g.

Ethanol yield: 70.62 g ÷ 250 g = 0.283 g ethanol/g sorghum.

Assume 1 bu of sorghum weighs 56 lb and has moisture content of 15 wt%. The mass of 1 bu of BRM sorghum in g is 56 lb × 0.85 × 454 g/lb = 21610 g.

Ethanol yield per bu of sorghum is 0.283 g/g × 21610 g/bu = 6104 g ethanol/bu or 2.04 gal/bu.

Similar calculations are performed for dewaxed and dewaxed/H_2SO_4-treated BRM sorghum. The ethanol yields are 2.59 gal/bu and 2.92 gal/bu, respectively.

Appendix A.2. Calculations of Potential Economic Benefits

Design basis: A plant to produce 50 million gallons ethanol per year.

Annual feedstock requirement for raw sorghum: 50×10^6 gal ÷ 2.04 gal/bu = 24.5×10^6 bu.

Similarly, for dewaxed sorghum and dewaxed/H_2SO_4-treated sorghum, the annual feedstock requirements are 19.3×10^6 bu and 17.1×10^6 bu.

For the dewaxed sorghum, the amount of wax that can be extracted is 0.003×21.6 kg/bu = 0.065 kg/bu. The total quantity of wax that can be extracted is 0.065 kg/bu × 19.3×10^6 bu = 1.251×10^6 kg or 1251 MT.

The value of the extracted wax is 1.251×10^6 kg × $6/kg = 7.5×10^6.

The saving on feedstock is (24.5×10^6 bu − 19.3×10^6 bu) × $3.10/bu = 16.1×10^6.

Similar calculations are performed for the dewaxed/H_2SO_4-treated sorghum and the calculated results are shown in Table 7.

References

1. Renewable Fuels Association. Ethanol Industry Outlook. 2017. Available online: http://www.ethanolrfa. org/wp-content/uploads/2017/02/Ethanol-Industry-Outlook-2017.pdf (accessed on 30 November 2017).
2. USDA. Feed Outlook Reports. Available online: http://usda.mannlib.cornell.edu/MannUsda/ viewDocumentInfo.do?documentID=1273 (accessed on 30 November 2017).
3. Nhuan, P.N.; Montanti, J.; Johnston, D.B. Sorghum as a renewable feedstock for production of fuels and industrial chemicals. *Bioengineering* **2016**, *3*, 75–91.
4. Hwang, K.T.; Cuppett, S.L.; Weller, C.L.; Hanna, H.A. Properties, composition and analysis of grain sorghum wax. *J. Am. Oil Chem. Soc.* **2002**, *79*, 521–527. [CrossRef]
5. Hwang, K.T.; Cuppert, S.L.; Weller, C.L.; Hanna, M.A. HPLC of grain sorghum wax classes highlighting separation of aldehydes from wax esters and steryl esters. *J. Sep. Sci.* **2002**, *25*, 619–623. [CrossRef]
6. Harron, A.F.; Powell, M.J.; Nunez, A.; Moreau, R.A. Analysis of sorghum wax and carnauba wax by reversed phase liquid chromatography mass spectrometry. *Ind. Crops Prod.* **2017**, *98*, 116–129. [CrossRef]
7. Steinle, J.V. CARNAUBA WAX an expedition to its source. *Ind. Eng. Chem.* **1936**, *28*, 1004–1008. [CrossRef]
8. Carnauba Wax. Available online: https://www.alibaba.com/product-detail/Carnauba-Wax_60433586534. html (accessed on 12 February 2018).
9. Carnauba Wax Market Analysis by Product. Available online: http://www.grandviewresearch.com/ industry-analysis/carnauba-wax-market (accessed on 12 February 2018).
10. Carnauba Wax Market Size Projected to Reach $334.9 Million by 2024. Available online: https://www. grandviewresearch.com/press-release/global-carnauba-wax-market (accessed on 12 February 2018).
11. Cai, D.; Chang, Z.; Wang, C.; Ren, W.; Wang, Z.; Qin, P.; Tan, T. Impact of sweet sorghum cuticular waxes (SSCW) on acetone-butanol-ethanol fermentation using *Clostridium acetobutylicum* ABE 1201. *Bioresour. Technol.* **2013**, *149*, 470–473. [CrossRef] [PubMed]
12. Wall, J.S.; Blessin, C.W. Composition and structure of sorghum grains. *Cereal Sci. Today* **1969**, *14*, 264–271.
13. Drapcho, C.M.; Nghiem, N.P.; Walker, T.H. *Biofuels Engineering Process Technology*; McGraw-Hill: New York, NY, USA, 2008; pp. 134–143.

14. Total Starch Assay Procedure, Megazyme. Available online: https://secure.megazyme.com/files/Booklet/K-TSTA_DATA.pdf (accessed on 12 February 2018).
15. Sluiter, A.; Hames, B.; Ruiz, R.; Scarlata, C.; Sluiter, J.; Templeton, D.; Crocker, D. *Determination of Structural Carbohydrates and Lignin in Biomass*; Technical Report NREL/TP-510-42618; National Renewable Energy Laboratory: Golden, CO, USA, 2011.
16. Hagerman, A.E.; Butler, L.G. Condensed tannin purification and characterization of tannin-associated proteins. *J. Agric. Food Chem.* **1980**, *28*, 947–952. [CrossRef] [PubMed]
17. Butler, L.G.; Riedl, D.J.; Lebryk, D.G.; Blytt, H.J. Interaction of proteins with sorghum tannin: Mechanism, specificity and significance. *J. Am. Oil Chem. Soc.* **1980**, *61*, 916–920. [CrossRef]

![fermentation logo] *fermentation*

MDPI

Review

Biological Production of 3-Hydroxypropionic Acid: An Update on the Current Status

Leonidas Matsakas *, Kateřina Hrůzová, Ulrika Rovaand Paul Christakopoulos

Biochemical Process Engineering, Division of Chemical Engineering, Department of Civil, Environmental and Natural Resources Engineering, Luleå University of Technology, SE-971 87 Luleå, Sweden; katerina.hruzova@ltu.se (K.H.); ulrika.rova@ltu.se (U.R.); paul.christakopoulos@ltu.se (P.C.)
* Correspondence: leonidas.matsakas@ltu.se; Tel.: +46-(0)-920-493043

Received: 22 January 2018; Accepted: 9 February 2018; Published: 13 February 2018

Abstract: The production of high added-value chemicals from renewable resources is a necessity in our attempts to switch to a more sustainable society. 3-Hydroxypropionic acid (3HP) is a promising molecule that can be used for the production of an important array of high added-value chemicals, such as 1,3-propanediol, acrylic acid, acrylamide, and bioplastics. Biological production of 3HP has been studied extensively, mainly from glycerol and glucose, which are both renewable resources. To enable conversion of these carbon sources to 3HP, extensive work has been performed to identify appropriate biochemical pathways and the enzymes that are involved in them. Novel enzymes have also been identified and expressed in host microorganisms to improve the production yields of 3HP. Various process configurations have also been proposed, resulting in improved conversion yields. The intense research efforts have resulted in the production of as much as 83.8 g/L 3HP from renewable carbon resources, and a system whereby 3-hydroxypropionitrile was converted to 3HP through whole-cell catalysis which resulted in 184.7 g/L 3HP. Although there are still challenges and difficulties that need to be addressed, the research results from the past four years have been an important step towards biological production of 3HP at the industrial level.

Keywords: 3-hydroxypropionic acid; metabolic engineering; building-block chemicals; glycerol; platform chemicals; *Klebsiella pneumoniae*; *Escherichia coli*; *Saccharomyces cerevisiae*

---

1. Introduction

The use of fossil resources for the production of fuels, chemicals, and materials has caused serious environmental problems, which—together with their imminent depletion—has made the establishment of renewable alternative production methods an important priority. For this reason, development of technologies for the establishment of biorefineries for the production of fuels, chemicals, and materials from renewable resources has been actively pursued to replace the use of fossil resources. Of the different molecules that are projected to be produced through biorefinery-based strategies, 3-hydroxypropionic acid (3HP) holds an important position.

3HP is one of the key building-block chemicals. In the list issued in 2004 and updated in 2010, the United States Department of Energy (DOE) recognized it as one of the 12 top building-block chemicals that can be produced from biomass, ranking it in third position among the molecules selected [1]. It contains two functional groups (a carboxyl group and a β-hydroxyl group), which makes it attractive to serve as an excellent versatile platform for the production of a variety of high added-value chemicals through chemical modification reactions [1]. The compounds that can be produced from 3HP include 1,3-propanediol, acrylic acid, acrylamide, acrylonitrile, propiolactone, malonic acid, homopolymers, and heteropolymers [2,3]. These compounds have a broad range of applications, and can be used for the production of adhesives, polymers, plastic packaging, fibers, cleaning agents, and resins. Until now, they have mainly been produced by the petrochemical industry from fossil resources. However, this is not sustainable, and there is an urgent need to switch to economically and ecologically sound production using renewable resources. Production of these substances from 3HP is a sustainable solution in itself, but it is also important that the production of 3HP itself should take place from renewable resources. For example, it has been estimated that production of acrylic acid from 3HP, which is otherwise produced from glycerol, may be 50% cheaper than petroleum-based acrylic acid production, with a 75% reduction in greenhouse gas emissions [4]. Much research is being devoted to the biotechnological production of 3HP from renewable resources. To achieve this goal, an intense research effort has been made (1) to identify novel biochemical pathways; and (2) regarding the metabolic engineering of microbial strains to transform them to cell factories capable of producing 3HP (the two main renewable carbon sources being glycerol and sugars). Moreover, attempts to increase 3HP production titers are also being made through improvements in cultivation techniques. These intense research activities have meant that a lot of research work has been published in the past four years. The aim of the current review is to provide an update on the progress that has occurred in the biotechnological production of 3HP after 2013, as an update to our previous review article on this matter [3].

2. Production of 3HP from Glycerol

The use of glycerol, which is a renewable resource, as starting material for the biotechnological production of 3HP, is a common strategy. It can be obtained as a by-product of biodiesel production. Biodiesel is produced from the transesterification of oils with a short-chain alcohol (mainly methanol) in the presence of a catalyst [5]. During biodiesel production, crude glycerol is generated as by-product in a ratio of about 100 kg glycerol to 1 ton of biodiesel produced [6]. As the worldwide production of biodiesel is tending to increase, the amounts of crude glycerol that are available will follow the same trend in the future. It is estimated that the production of glycerol will reach 4.2×10^9 L in 2020, and that the world glycerol market will have a value of about 2.52 billion US dollars (USD) [7,8]. Moreover, the oversupply of glycerol leads to a decrease in its price [9]. During 2013, the prices of refined glycerol varied from 900 USD/ton to 965 USD/ton, with the prices of the unrefined crude glycerol being as low as 240 USD/ton [7]. The high quantities of crude glycerol that are produced require its treatment or further use, which, together with the low price, creates an ideal situation for the incorporation of crude glycerol into the biotechnological production of a variety of fuels and chemicals [8,9]. Moreover, upgrading of biodiesel-derived glycerol to high added-value products will boost the economy of the biodiesel production sector [10]. Conversion of glycerol to 3HP is one of the available options for the utilization of glycerol, and much research effort is being made towards this conversion. Several bacterial strains, such as *Klebsiella* sp., are capable of natural uptake of glycerol through oxidative or reductive pathways. In the oxidative pathway, this leads to the formation of pyruvate through the Embden–Meyerhof–Parnas pathway, whereas the reductive pathway results in the formation of 1,3-propanediol [11]. Genetic manipulation of the natural glycerol consuming strains can result in the conversion of glycerol to 3HP, through 3-hydroxypropionaldehyde as an intermediate. To achieve this conversion, the genes of either the *dha* operon or the *pdu* operon are commonly used, as summarized in the following sections.

2.1. Production Using Genes That Are Part of the Dha Operon

The naturally occurring *dha* operon encodes the necessary enzymes for the conversion of glycerol to 1,3-propanediol. A detailed description of the genes in the *dha* operon and their action is given in our previous review article [3]. Briefly, the operon mainly consists of genes encoding (1) a glycerol dehydratase, which catalyzes the conversion of glycerol to 3-hydroxypropionaldehyde; (2) a reactivase of the glycerol dehydratase; and (3) a 1,3-propanediol oxidoreductase, which catalyzes the conversion of 3-hydroxypropionaldehyde to 1,3-propanediol, with some structural differences in different microorganisms. Conversion of glycerol to 3HP using the *dha* operon is performed in two stages, whereby glycerol is initially converted to 3-hydroxypropionaldehyde through the action of the enzyme glycerol dehydratase, followed by conversion of the 3-hydroxypropionaldehyde to 3HP by the action of an aldehyde dehydrogenase (ald) enzyme (Figure 1). A strategy that is commonly employed is to use a microorganism that already possess the *dha* operon, and genetically modify it to produce 3HP. In that case, the gene encoding the 1,3-propanediol oxidoreductase should be deleted or underexpressed if the co-production of 1,3-propanediol is not wanted, but special care should be taken to maintain a balance between NADH and NAD^+. Regeneration of NAD^+ is important, as it is required for the action of the ald enzyme. NAD^+ can be regenerated in the electron transport chain, which requires increased aeration, but the presence of oxygen inhibits the synthesis of coenzyme B_{12}, which is also involved in the process [12]. Moreover, as will be discussed later, oxygen also inactivates the glycerol dehydratase. Regeneration of NAD^+ can also take place during the conversion of 3-hydroxypropionaldehyde to 1,3-propanediol by the action of the 1,3-propanediol oxidoreductase, which consumes NADH [13]. If the 1,3-propanediol oxidoreductase gene is knocked out, another pathway for NAD^+ should be found in order to maintain the balance between NADH and NAD^+, such as lactate or acetate formation [12]. On the other hand, co-production of 3HP and 1,3-propanediol can also be desirable, as 1,3-propanediol also has significant applications and these two molecules can be separated easily, due to the presence of different functional groups on the molecules [14].

Another consideration when the *dha* operon is used is that the first step of the reaction requires coenzyme B_{12}, which is produced naturally de novo by some bacterial strains, such as *Klebsiella pneumoniae* [15,16], whereas other bacteria, such as *Escherichia coli*, cannot produce it de novo [16]. When the host microorganism is not capable of producing this coenzyme, external addition of it should be included in the cultivation—which will affect the cost of the process. An effort is therefore being made to create new bacterial strains that are capable of producing coenzyme B_{12}. Another consideration when using the *dha* operon is the inactivation of the glycerol dehydratase during the catalysis of glycerol conversion [17,18], so proper expression of the glycerol dehydratase reactivase is important to maintain the activity of glycerol dehydratase and, in turn, 3HP production. Glycerol dehydratase can also be inactivated by oxygen [19], which on the other hand, is necessary for efficient regeneration of the NAD^+ [20] that is required for the action of the ald enzymes.

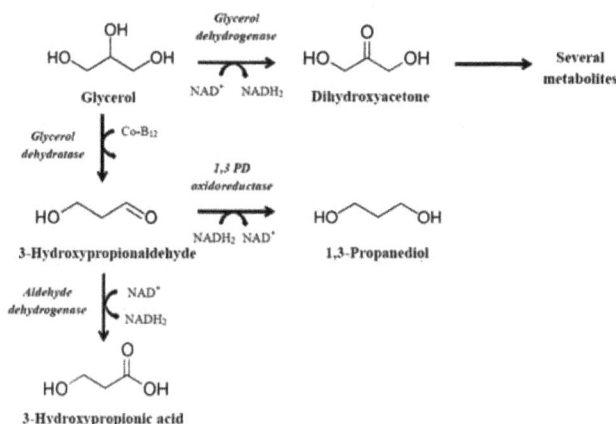

Figure 1. Conversion of glycerol to 3-hydroxypropionic acid (3HP) employing genes from the *dha* operon.

2.2. Production Using Genes That Are Part of the Pdu Operon

The second operon that can be used for the construction of a strain capable of converting glycerol to 3HP is the *pdu* operon. In nature, the *pdu* operon is required for microbial growth on 1,2-propanediol and it can be found in *Salmonella* and *Lactobacillus* species [21,22]. First, 1,2-propanediol is converted to propionaldehyde by the action of an AdoCbl-dependent diol dehydratase; this is further converted to propionic acid and propanol by the action of the enzymes CoA-dependent aldehyde dehydrogenase, phosphotransacylase, propionate kinase, and alcohol dehydrogenase [22]. Regarding the conversion of glycerol to 3HP by the enzymes encoded in the *pdu* operon, more intermediate steps are required than when using the *dha* operon. Initially, glycerol is converted to 3-hydroxypropionaldehyde by the action of the diol dehydratase, which is followed by its transformation to 3-hydroxypropionyl-CoA by the action of propionaldehyde dehydrogenase. The third step consists of phosphorylation of 3-hydroxypropionyl-CoA to 3-hydroxypropionyl phosphate by phosphate propanoyltransferase. Finally, dephosphorylation of 3-hydroxypropionyl phosphate, by the enzyme propionate kinase, leads to the formation of 3HP (Figure 2).

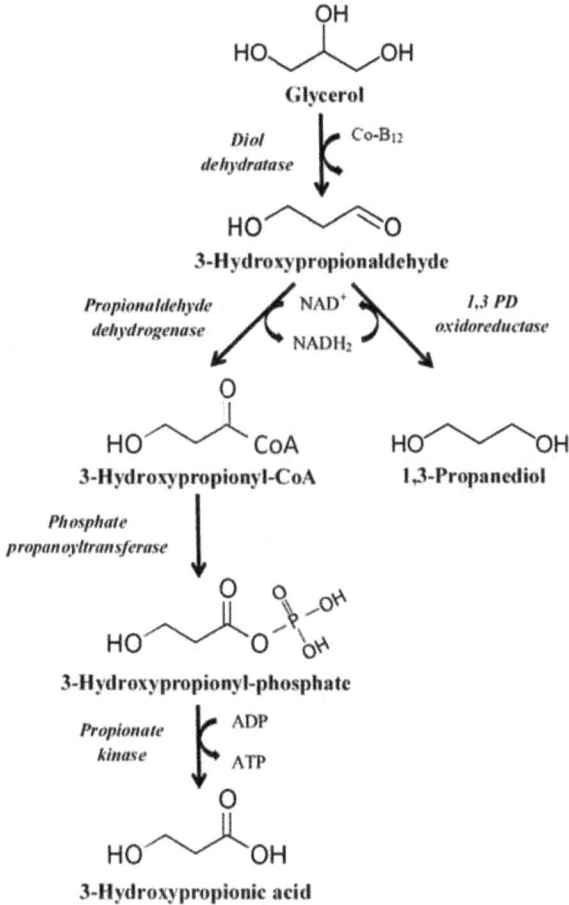

Figure 2. Conversion of glycerol to 3HP employing genes from the *pdu* operon.

Similarly to the *dha* pathway, the use of the *pdu* pathway presents some challenges that should be taken into account when intending to use it for industrial production of 3HP. As NAD$^+$ is required during the second step of the pathway (conversion of 3-hydroxypropionaldehyde to 3-hydroxypropionyl-CoA), an adequate supply of NAD$^+$ is necessary. Similar strategies and considerations as described before apply here also, with the co-production of 1,3-propanediol serving as a promising solution, as the action of 1,3-propanediol oxidoreductase requires NADH, and therefore, NAD$^+$ is regenerated [23,24], maintaining the equilibrium between NAD$^+$ and NADH in the process.

2.3. Engineering of K. pneumoniae Cells for the Conversion of Glycerol to 3HP

K. pneumoniae is an important candidate for use in the conversion of glycerol to 3HP. *K. pneumoniae* can naturally produce the required coenzyme B_{12} de novo, which is important for the economics of the process, as there is no need for external addition of coenzyme B_{12}. Moreover, the *dha* operon is endogenous in *K. pneumoniae*, thus minimizing the transfer of required genes when constructing a strain capable of producing 3HP, and has excellent glycerol fermentation capability [25]. On the other hand, as 1,3-propanediol is produced through the *dha* operon, the genes encoding the 1,3-propanediol oxidoreductase should be deleted, unless co-production of 1,3-propanediol is desired. On the downside, use of *K. pneumoniae* can potentially give rise to public health concerns. Bacteria of the *Klebsiella* genus can cause human nosocomial infection, due to their ability to spread rapidly in hospital environments. *K. pneumoniae*, in particular, is the most important of the *Klebsiella* genus, from a medical point of view, as it can cause a broad variety of infections, such as urinary tract infections, soft tissue infections, septicemias, and pneumonia in hospital environments [26].

As was discussed before, the balance of the cofactors NADH and NAD^+ is important during the conversion of glycerol to 3HP. A strategy to maintain this balance is the co-production of 3HP and 1,3-propanediol. For example, Su et al. [27] constructed a strain through the heterologous expression of *DhaS*, a putative ald from *Bacillus subtilis*, which showed higher specificity toward 3-hydroxypropionic acid than toward other aldehydes (propionaldehyde, benzaldehyde, valeraldehyde, butyraldehyde, and acetaldehyde). The recombinant *K. pneumoniae* strain was capable of producing 3HP at 18.0 g/L, and 1,3-propanediol at 27 g/L in 24 h under non-optimized bioreactor conditions. During the cultivation, a significant amount of lactic acid (36.1 g/L) was also produced (Table 1). The authors suggested that some possible reasons for the low 3HP concentration might be the competition for NAD^+ between 3HP and biomass formation, the toxicity of 3HP toward the host cells, and the non-optimal activity of the ald toward 3-hydroxypropionaldehyde. It is obvious from the above that an adequate supply of NAD^+ is necessary to achieve a high degree of 3HP production, and the discovery of more efficient alds is also important.

Table 1. Results of 3HP production from glycerol using *K. pneumoniae*.

Genes Transferred	Culture Conditions	Concentration (g/L)	Productivity (g/L·h)	Ref.
dhaS gene from *B. subtilis*	Fed-batch bioreactor 5 L, 3 L, pH 7, 37 °C, 1.5 vvm, 400 rpm	18.0	0.77	[27]
aldH gene from *E. coli*	Fed-batch bioreactor, 5 L, 2.8 L, pH 6.8–7.0, 37 °C, microaeration 1.5 vvm, 400 rpm	48.9	1.75	[28]
Overexpression of *kgsadh*, *dhaB* and *gdrAB*, deletion of *ldhA*, *frdA*, and *adhE*	Fed-batch bioreactor 1.5 L, 1 L, pH 7.5, 37 °C, 1 vvm, 400 rpm	43.0	0.90	[12]
PuuC overexpression, deletion of *ldh1*, *ldh2*, and *pta*	Fed-batch bioreactor 5 L, 3 L, pH 7, 37 °C, 1.5 vvm, 400 rpm	83.8	1.16	[29]
aldH gene from *E. coli*	Flasks 250 mL, 100 mL, 37 °C, microaeration, 150 rpm	0.9	0.04	[25]
-	Co-cultivation with *Gluconobacter oxydans* in fed-batch bioreactor, 7 L, 4.0 L, 1st step: pH 7, 37 °C, 0.2 vvm, 150 rpm 2nd step: pH 5.5, 28 °C, 0.5 vvm, 600 rpm	60.5	1.12	[30]

The strategy of co-production was also followed by Huang et al. [28], who expressed the *aldH* gene from *E. coli* in *K. pneumoniae* cells. The authors suggested that the relative amounts of 3HP and 1,3-propanediol could be controlled by controlling the aeration levels in the culture. Cell growth and production of 3HP were enhanced by increasing the aeration rate, but 1,3-propanediol production was reduced. Under fully aerobic conditions, however, the *dha* operon was repressed—resulting in no production of either 3HP or 1,3-propanediol. The highest 3HP concentration (48.9 g/L) was reached under microaerobic conditions (1.5 vvm aeration, which resulted in decrease of the dissolved oxygen from 100% to 0% in 2.7 h of culture), with the simultaneous production of 25.3 g/L 1,3-propanediol after 28 h of cultivation in a fed-batch bioreactor. The overall yield of 1,3-propanediol and 3HP was 0.66 mol/mol. The authors also demonstrated that the formation of other by-products was also affected by the level of aeration under micro-anaerobic conditions, with the production of lactic acid and acetic acid being enhanced, whereas ethanol and succinic acid production was reduced by increasing the aeration level. Formate production initially increased with increasing levels of aeriation, followed by a sharp decrease when aeration was higher than 0.6 vvm. Ko et al. [12] also attempted to control the production of other metabolic by-products, especially of acetic acid, with the aim of improving co-production of 3HP and 1,3-propanediol from the recombinant strain of *K. pneumoniae* (J2B), which overexpressed the ald gene (encoding alpha-ketoglutaric semialdehyde dehydrogenase—*kgsadh*). The methods that they evaluated for acetate reduction were reduction of the glycerol assimilation through the glycolytic pathway, increasing the glycerol flow towards 3HP and 1,3-propanediol formation, and finally controlling the aeration levels. To improve the co-production, the authors evaluated the deletion and overexpression of several genes. The best results were obtained when the genes encoding lactate dehydrogenase (*ldhA*), succinate dehydrogenase (*frdA*), and alcohol dehydrogenase (*adhE*) were deleted, which—together with the overexpression of *dhaB* and *gdrAB*—resulted in 3HP at 43 g/L and 1,3-propanediol at 21 g/L during fed-batch bioreactor cultivation, with an overall yield of 0.49 mol/mol. Although these genetic manipulations reduced the amount of acetate, a considerable amount of acetate (>150 mM) had accumulated in the bioreactor by the end of the cultivation.

In another study conducted by Li et al. [29], a systematic optimization of glycerol metabolism took place in order to improve the 3HP production titers. During this work, different promoters for the overexpression of *PuuC* (a native ald of *K. pneumoniae*), which is a key enzyme for 3HP formation, were investigated. Among the promoters tested (*tac* and *lac*), the IPTG-induced *tac* was found to be the most efficient for overexpression of *PuuC*. Moreover, 3HP production significantly increased when the synthesis of lactic acid and acetic acid was blocked. Finally, optimization of cultivation parameters, such as aeration (microaerobic conditions), pH (7.0), and IPTG concentration (0.02 mM) improved the production of 3HP even further, resulting in a concentration of 83.8 g/L with a yield 52 g/g after 72 h of cultivation. To the best of our knowledge, this concentration is the highest reported from glycerol. Although addition of IPTG is not optimal from an economic point of view, the amount required during this work was relatively low compared to what is commonly used (0.5–2 mM IPTG was used for *tac*-driven gene expression in *E. coli*, and up to 5 mM IPTG was used for *Zymomonas mobilis* or *Pseudomonas putida*) [29]. The high concentration of 3HP demonstrated is an important step towards industrial application.

The most frequent approach for the genetic engineering of novel strains is plasmid insertion (of the required genes) into the host strain. However, as strains containing plasmids are not genetically stable and they require inducers and antibiotics to maintain the selection pressure during cultivation, Wang and Tian [25] tried a different approach for construction of the host strain. More specifically, they constructed a plasmid-free *K. pneumoniae* strain through chromosomal engineering by replacing the IS1 region in the chromosome with the AD DNA cassette containing the *aldH* gene from *E. coli* through homologous recombination. This strain was able to produce 3HP at 0.9 g/L (Table 1) in flask cultures when glycerol, at 40 g/L, was added. Although the concentration of the 3HP produced was low, this work demonstrated a new approach for the construction of host strains without any need for inducers and antibiotics, which could be very useful for the development of new strains in the future.

Finally, a totally different approach to production of 3HP from glycerol was studied by Zhao et al. [30]. In their process, they used a two-step approach, where glycerol was first converted to 1,3-propanediol by *K. pneumoniae*, followed by conversion of 1,3-propanediol to 3HP by *Gluconobacter oxydans*—a bacterium that incompletely oxidizes a wide range of ketones, organic acids, and aldehydes. The final concentration of 3HP was 60.5 g/L, and the conversion rate of glycerol to 3HP was 0.5 g/g. Moreover, it was the first time that acrylic acid production was reported as a by-product of 3HP production from 1,3-propanediol, at a concentration of approximately 1 g/L—another high added-value chemical with several applications, such as plastics, adhesives, and coatings.

2.4. Engineering of E. coli Cells for the Conversion of Glycerol to 3HP

E. coli is commonly used as a host microorganism for genetic modifications, and there is a wide range of commercial genetic tools available, making it easy to handle with efficient control of gene transfer. In addition, *E. coli* has a large number of alds in its genome—genes that could be studied and overexpressed for the conversion of glycerol to 3HP. However, the disadvantage of *E. coli* is that it is not able to produce coenzyme B$_{12}$, leading to the necessity of external addition of this expensive compound, thus affecting process economics and industrial applications.

As when *K. pneumoniae* is used as the host strain, when using *E. coli*, one common strategy is to use the genes of the *dha* operon. As has been discussed, glycerol dehydratase undergoes rapid inactivation, which can result in shutdown of 3HP production. Niu et al. [31] demonstrated that when the glycerol dehydratase reactivase gene was cloned together with the glycerol dehydratase gene, the 3HP concentration increased fivefold, due to the prevention of glycerol dehydratase inactivation. In an attempt to increase the 3HP production yield, the authors also expressed the gene encoding the NAD$^+$-regenerating enzyme, glycerol-3-phosphate dehydrogenase (Table 2). However, the concentration of 3HP was reduced, with increased production of malic acid, due to the lack of NAD$^+$. Enhancement of 3HP production by regulation of glycerol metabolism and minimizing of by-product formation was attempted by Jung et al. [4]. During their work, they tried to eliminate the formation of major by-products, such as acetate and 1,3-propanediol, and to increase the metabolic flow of glycerol towards 3HP by upregulating the glycerol kinase (*glpK*) and the glycerol facilitator (*glpF*), and by deleting the regulatory factor that repressed the use of glycerol (*glpR*). After these modifications, the generation of by-products was minimized, and the uptake of glycerol was improved, resulting in a production of 3HP as high as 42.1 g/L. The average yield was 0.268 g/g.

When aiming to produce 3HP through the 3-hydroxypropionaldehyde intermediate, proper balancing of the steps should be attempted, to avoid accumulation of the 3-hydroxypropionaldehyde. 3-Hydroxypropionaldehyde is very toxic for microorganisms, even at concentrations of 15–30 mM, and its accumulation can result in inhibition of 3HP production, with concentrations as low as 10 mM or even lower having a significant negative effect on growth and enzyme activity [32,33]. One solution to the problem could be the selection of an efficient aldehyde dehydrogenase that would rapidly act on 3-hydroxypropionaldehyde and convert it to 3HP. Based on this strategy, Chu et al. [34] tried 17 candidate aldehyde dehydrogenases for their activity against 3-hydroxypropionaldehyde, with the *gabD4* from *Cupriavidus necator* turning out to be the most effective one. In an attempt to further improve the aldehyde dehydrogenase selected, the authors performed site-directed and saturation mutagenesis, based on homologous modeling. The mutant enzyme obtained had 1.4-fold higher activity compared to the wild type one, and a high 3HP production of 71.9 g/L (with a productivity of 1.8 g/L·h) was achieved in fed-batch bioreactor culture, which, to the best of our knowledge, is the highest reported concentration reported with *E. coli* as a host growing on glycerol.

Another strategy to avoid 3-hydroxypropionaldehyde accumulation is the use of promoters with different strength in controlling the expression of the enzymes [3], although the number of genetic elements can limit the control over gene expression [33]. The use of the in silico design tool "UTR Designer" has also been proposed; it can provide precise predictions of the translation initiation efficiency [33]. This tool was employed by Lim et al. [33] in their effort to prevent the accumulation of 3-hydroxypropionaldehyde by fine-tuning the expression levels of aldehyde dehydrogenase and glycerol dehydratase. Moreover, they deleted the by-product formation genes, *yghD* and *ackA-pta*, in order to improve metabolic flow towards the formation of 3HP. During flask culture, 3HP formation reached 17.9 g/L with a yield of 0.61 g/g. Following on from these results, the authors tried a fed-batch cultivation with the addition of glucose together with the glycerol. Under these conditions, the production of 3HP increased to 40.5 g/L with a yield of 0.97 g/g. This increase in the yield of 3HP formation could be explained by the better carbon flow of glycerol towards 3HP formation, as glucose could cover the needs of cell growth. A different approach for balancing the pathway enzymes to prevent 3-hydroxypropionaldehyde accumulation, and in turn, improve the 3HP production yields, was used by Sankaranarayanan et al. [35]. In their work, they used a synthetic regulatory cassette comprised of varying-strength promoters and bicistronic ribosome-binding sites (RBSs) to control the expression of the genes. Fine-tuning of the levels of expression between the two genes could result in no secretion of 3-hydroxypropionaldehyde, which was achieved when aldehyde dehydrogenase had an expression that was 8-fold higher than that of glycerol dehydratase. This strategy resulted in the engineering of an *E. coli* strain capable of producing up to 56.4 g/L 3HP in a fed-batch bioreactor, with the addition of glucose together with glycerol. The addition of glucose as a co-substrate was also found to improve the activity of the aldehyde dehydrogenase gene, and in turn, improve the 3HP production yields [36]. Niu et al. [36] reported a 3.5-fold increase in the activity of the enzyme, which improved the 3HP production in flasks from 3.39 g/L (control—no addition of glucose) to 6.80 g/L. Optimization of the glucose concentration and feeding strategy improved the 3HP production to up to 17.2 g/L during fed-batch cultivation. The authors also suggested that addition of glucose reduced the imbalance between the activities of glycerol dehydratase and aldehyde dehydrogenase, as it led to an increase in the activity of aldehyde dehydrogenase.

Table 2. Results of 3HP production from glycerol using *E. coli*.

Genes Transferred	Culture Conditions	Concentration (g/L)	Productivity (g/L·h)	Ref.
dhaB and *gdrAB* from *K. pneumoniae* and *kgsadh* from *A. brasilense*	Flasks 250 mL, 100 mL, 37 °C, 200 rpm	5.1	n.a.	[31]
dhaB and *gdrAB* from *K. pneumoniae*, *aldH* from *E. coli*, overexpression of *glpF*, deletion of *ackA-pta*, *yqhD*, and *glpR*	Fed-batch bioreactor 5 L, 2 L, pH 7, 35 °C, 1 vvm, 500 rpm	42.1	1.32	[4]
dhaB and *gdrAB* from *K. pneumoniae*, *gabD4* from *C. necator* with side-directed mutagenesis, deletion of *yghD* and *ackA-pta*	Fed-batch bioreactor 5 L, 2 L, pH 7, 35 °C, 1 vvm, 500 rpm	71.9	1.8	[34]
UTR-engineered *dhaB*, *gdrAB* from *K. pneumoniae* and *kgsadh* from *A. brasilense*, deletion of *yghD* and *ackA-pta*	Fed-batch bioreactor 5 L with the addition of glucose, 2 L, pH 7, 37 °C, 1 vvm, 500 rpm	40.5	1.35	[33]
dhaB and *gdrAB* from *K. pneumoniae* and *kgsadh* from *A. brasilense*	Fed-batch bioreactor 1.5 L with the addition of glucose, pH 7, 37 °C, 1 vvm, 650 rpm	56.4	1.18	[35]
dhaB and *gdrAB* from *K. pneumoniae* and *kgsadh* from *A. brasilense*	Fed-batch bioreactor 5 L with the addition of glucose, 3 L, pH 6.5, 37 °C, 450 rpm	17.2	n.a.	[36]
dhaB and *gdrAB* from *K. pneumoniae* and *kgsadh* from *A. brasilense*	Fed-batch bioreactor 1.5 L, 1 L, pH 7, 37 °C, 0.5 vvm, 650 rpm	41.5	0.86	[37]
dhaB and *gdrAB* from *K. pneumoniae*, *AraE* from *A. brasilense*, conditional repression of *gapA*, deletion of *yqhD*	Flasks, 300 mL, 37 °C, 150 rpm	6.06	0.13	[38]
dhaB and *gdrAB* from *K. pneumoniae*, *AraE* from *A. brasilense*, *pdu* from *K. pneumoniae*	Flasks 250 mL, 60 mL, pH 6–7, 100–250 rpm	5.05	0.105	[39]

n.a.: not available.

Not only the presence of 3-hydroxypropionaldehyde, but also high concentrations of 3HP can inhibit the growth of the host microorganisms, thus hindering the production of 3HP. Aiming to identify a 3HP-tolerant *E. coli* strain, Sankaranarayanan et al. [37] studied nine acid-tolerant strains that efficiently produced various organic acids at high titers. Construction of the 3HP producing strains was achieved by expressing the *dha* operon, and an ald from *Azospirillum brasilense*. Of all the strains tested, two showed a high degree of growth in the presence of 25 g/L 3HP, and one of them, *E. coli* W, outperformed the rest, and produced 41.5 g/L 3HP with a yield of 0.31 g/g. During this work, it was found that there are significant differences in 3HP tolerance among *E. coli* strains, and this should be taken into account when selecting an appropriate host microorganism.

To improve the metabolic flow of glycerol towards 3HP, other genes of glycerol catabolism are often deleted, which might result in strains with poor growth [38]. To avoid this and at the same time improve the metabolic flow towards 3HP, Tsuruno et al. [38] suggested the use of a metabolic toggle switch (MTS), and tested the conditional repression of the following genes: *glpK* (encoding glycerol kinase), *tpiA* (encoding triosephosphate isomerase), and *gapA* (encoding glyceraldehyde-3-phosphate dehydrogenase). After testing the three different strains for their efficiency in 3HP production, it was found that only the strain with the MTS for *gapA* improved the production of 3HP to 4.88 g/L. Deletion of the gene *yqhD* (whose protein product is responsible for the conversion of 3-hydroxypropionaldehyde to 1,3-propanediol) further improved the concentration of 3HP to 6.06 g/L, with a yield of 0.515 mol/mol.

Finally, another strategy for genetically modifying *E. coli* was proposed by Honjo et al. [39], who used a dual synthetic pathway for the construction of *E. coli* strains. More specifically, they transferred the genes from both the *dha* and the *pdu* operons for the construction of one strain, and compared with the results obtained using a strain with only the genes from the *dha* operon. The strain with both pathways produced 3HP at 5.05 g/L, whereas the strain with only the *dha* operon produced only 2.98 g/L 3HP. The yield was 0.54 mol/mol. When the strain only had the *pdu* operon, it produced 1.41 g/L, thus underpinning the synergistic action of the two pathways in the conversion of glycerol to 3HP.

2.5. Other Microorganisms

It is clear that most of the research is focused mainly on either *E. coli* or *K. pneumoniae* as host microorganism for the construction of a cell factory capable of producing 3HP from glycerol. However, there are other microorganisms that could be suitable and efficient cell factories for efficient production of 3HP. For example, *Lactobacillus reuteri* is a good candidate, due to its ability to naturally produce the coenzyme B_{12} and to its high acid tolerance—both of which are required when producing 3HP. Moreover, the *pdu* operon is endogenous in that specific species [40], minimizing the need for gene transfer. The use of a recombinant *L. reuteri* strain with a mutation in the catabolite repression element (CRE) was investigated by Dishisha et al. [23], to improve the metabolic flux of glycerol to 3HP and 1,3-propanediol. During the flux analysis for the different steps of the *pdu* operon, it was found that the glycerol dehydration to 3-hydroxypropionaldehyde was ten times faster than the subsequent oxidation and reduction of 3-hydroxypropionaldehyde to 1,3-propanediol and 3HP. Thus, establishment of an optimal feeding rate of glycerol was crucial to avoid 3-hydroxypropionaldehyde accumulation and direct the flux of glycerol towards 3HP and 1,3-propanediol formation. The final titers were 10.6 g/L 3HP and 9.0 g/L 1,3-propanediol under anaerobic conditions in the fed-batch bioreactor when resting cells of *L. reuteri* were used (Table 3).

The same recombinant strain of *L. reuteri* was also used for the preparation of crosslinked, cryostructured monoliths that could be used as a biocatalyst for the conversion of glycerol to 3-hydroxypropionaldehyde, 3HP, and 1,3-propanediol [14]. Different crosslinkers were tested for the preparation of monolith columns of resting *L. reuteri* cells, with only the mixture of synthetic macromolecular structures of activated polyethyleneimine and modified polyvinyl alcohol (Cryo-PEI/PVA) demonstrating enhanced biocatalytic activity, mechanical stability, and sustained viability. Under optimal conditions, 3.3 g/L 3HP was produced during fed-batch feeding of the immobilized cells. Finally, a two-step process involving the cultivation of *L. reuteri* and *G. oxydans* has also been proposed [41]. More specifically, in the first step, the anaerobic cultivation of *L. reuteri* in fed-batch mode resulted in the production of equimolar quantities of 3HP and 1,3-propanediol, whereas in the second step, the 1,3-propanediol in the cell-free supernatant was selectively oxidized to 3HP by *G. oxydans* under aerobic batch cultivation. At the end of the first step, 14 g/L 3HP and 12 g/L 1,3-propanediol were produced, with the corresponding conversion yields being 0.48 g/g and 0.42 g/g, respectively. Finally, the oxidation in the second step resulted in the quantitative conversion of 1,3-propanediol to 3HP, with a final 3HP concentration of 23.6 g/L and an overall conversion yield of glycerol to 3HP of approximately 1 mol/mol.

Another host strain that has been used is *Bacillus subtilis*, which shows high growth rates, and is a non-pathogenic microorganism classified as GRAS (generally recognized as safe) [42]. The first attempt to use this host microorganism for 3HP was described by Kalantari et al. [42]. The constructed strain performed well in shaker flasks, producing up to 10 g/L with average yield 0.79 g/g, and showing good tolerance towards the 3HP produced. One drawback of the use of this strain as host is the lack of any native ability to produce the coenzyme B_{12} (as with *E. coli*); the authors suggested that this can be solved by transferring the necessary genes from *Bacillus megaterium*. This work demonstrated that *Bacillus* strains can also be used as cell factories for 3HP production, and more work using these cells will probably be conducted in the future.

Table 3. Results of 3HP production from glycerol using other host microorganisms.

Genes Transferred	Microorganism	Culture Conditions	Concentration (g/L)	Productivity (g/L·h)	Ref.
Mutation in the CRE upstream of the *pdu* operon	*L. reuteri*	Fed-batch bioreactor 3 L with resting cells, 1 L, pH 7, 37 °C, anaerobic, 500 rpm	10.6	1.08	[23]
Mutation in the CRE upstream of the *pdu* operon	*L. reuteri*	Immobilized cells with fed-batch feeding, 12 mL, 37 °C, anaerobic, 100 rpm	3.3	0.09	[14]
-	*L. reuteri* and *G. oxydans*	Two-step cultivation: (1) anaerobic fed-batch bioreactor 3 L with *L. reuteri* resting cells, 1 L, pH 5.5, 37 °C (2) aerobic batch system 3 L with *G. oxydans* resting cells, 1 L, pH 5.5, 28 °C, 0.33 vvm, 800 rpm	23.6	n.a.	[41]
dhaB, *gdrAB*, and *puuC* from *K. pneumoniae*, deletion of *glpK*	*B. subtilis*	Flasks with the addition of glucose, 37 °C, 200 rpm	10	n.a.	[42]

n.a.: not available.

3. Production of 3HP from Sugars

Glucose and other sugars (such as xylose) are another source of renewable raw materials for the cultivation of microorganisms when aiming to produce 3HP. Commonly used sources of sugar include sucrose from sugar beet and sugar cane, and glucose from the hydrolysis of corn starch. Due to the fact that these sources of sugar can also serve as food and animal feed, the use of sugars from lignocellulose presents an attractive alternative. Sources of lignocellulosic biomass include agricultural wastes and by-products, forest biomass, energy crops, and municipal waste [43–45]. The use of lignocellulosic materials as a source of fermentable sugars requires a pretreatment step (to disrupt the rigid structure of lignocellulosic biomass), followed by enzymatic hydrolysis of the insoluble carbohydrates in order to release soluble sugars. Several pretreatment methods have been proposed in the literature, which can be classified as physical, physicochemical, chemical, and biological [46]. The different pretreatment methods have shown varied efficiency against lignocellulosic biomass, with the source of the biomass affecting the chosen pretreatment parameters.

Regarding the use of sugars (mainly glucose) as carbon source for the production of 3HP, several different pathways have been proposed. Different research groups have made significant progress in this field by intensively studying some of these pathways. On the other hand, some other predicted pathways have turned out to not be efficient enough when transferred to host microorganisms.

3.1. Pathways for the Conversion of Sugars (Mainly Glucose) to 3HP

3.1.1. The Malonyl-CoA Pathway

To the best of our knowledge, malonyl-CoA pathway is currently one of the most investigated pathways for 3HP production. In this case, glucose is transformed through glycolysis to acetyl-CoA, which is then converted to malonyl-CoA by acetyl-CoA carboxylase. Malonyl-CoA reductase converts malonyl-CoA to 3HP thought a malonate semialdehyde intermediate [47] (Figure 3).

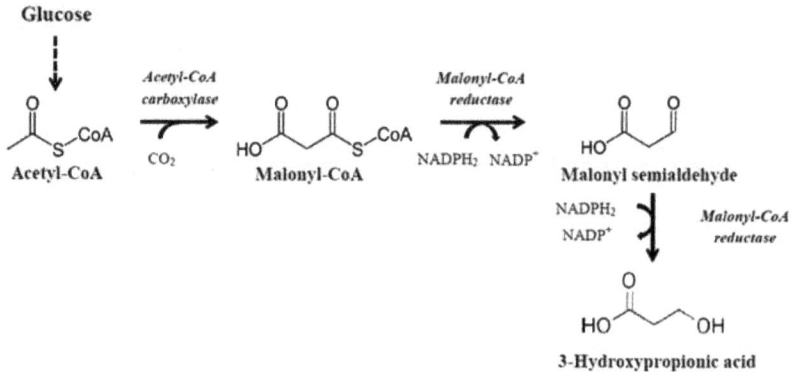

Figure 3. Conversion of glucose to 3HP through the malonyl-CoA pathway.

3.1.2. The β-Alanine Pathway

The main precursor of this pathway is aspartate, which is produced from fumarate (an intermediate of the citrate cycle). For this reason, it is advantageous to use a strain capable of overproduction of fumarate and of expressing a highly active aspartase, the enzyme responsible for the transformation of fumarate to aspartate. The next step is the transformation of aspartate to β-alanine by the action of aspartate-α-decarboxylase, followed by cleavage of the amino group of β-alanine by β-alanine-pyruvate aminotransferase or γ-aminobutyrate transaminase, resulting in malonic semialdehyde formation [48]. The last step is the conversion of malonic semialdehyde to 3HP by the action of malonic semialdehyde reductase, 3-hydroxypropionate dehydrogenase, or 3-hydroxyisobutyrate dehydrogenase [49] (Figure 4). Theoretically, the yields achieved through the β-alanine intermediate pathway should be higher than the yields obtained through the malonic intermediate pathway, as the malonic pathway is highly oxygen-dependent and requires high levels of ATP for acetyl-CoA synthesis [49].

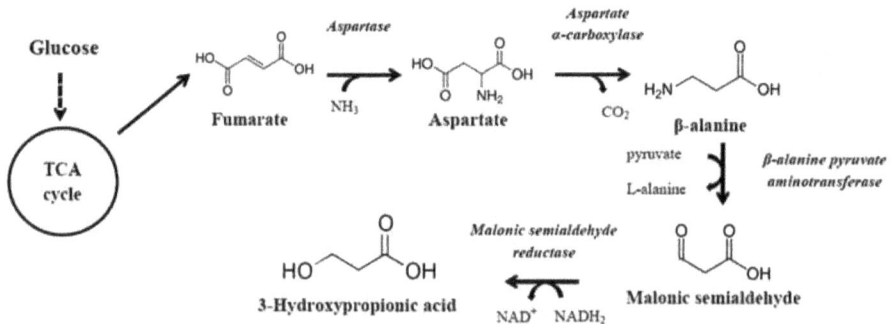

Figure 4. Conversion of glucose to 3HP through the β-alanine pathway.

3.1.3. The Propionyl-CoA Pathway

Another option is the conversion of glucose to 3HP through the intermediate propionyl-CoA, which was initially reported in the yeast *Candida rugosa* strain NPA-1 [50]. In this pathway, succinic acid can be produced from glucose and converted to propionic acid by the action of succinate decarboxylase, followed by conversion of propionate to propionyl-CoA by propionyl-CoA synthetase. Propionyl-CoA is then converted to acryloyl-CoA by the enzyme propionyl-CoA dehydrogenase. In the next step, the acryloyl-CoA is converted to 3-hydroxypropionyl-CoA by 3-hydroxypropionyl-CoA dehydratase, and finally, to 3HP through the action of the enzyme propionate-CoA transferase or the enzyme 3-hydroxyisobutyryl-CoA hydrolase [47,51] (Figure 5).

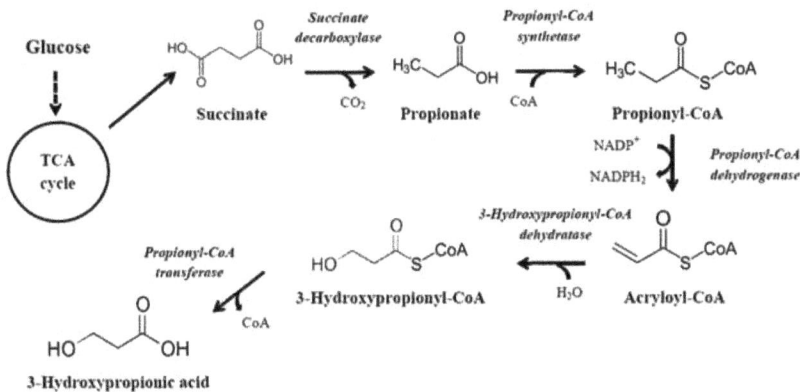

Figure 5. Conversion of glucose to 3HP through the propionyl-CoA pathway.

3.1.4. The Glycerate Pathway

The glycerate pathway is another option that has been proposed for the conversion of glucose to 3HP, but to the best of our knowledge, this pathway has not yet been constructed in a host microorganism. The pathway was proposed by Burgard and Van Dien [52]. In this proposed pathway, glucose is converted initially to 3-phosphate-glycerate, followed by its conversion to glycerate and malonate semialdehyde by the action of the enzymes 3-phosphate-glycerate phosphatase and glycerate dehydratase, respectively. Finally, malonate semialdehyde is converted to 3HP by 3HP dehydrogenase.

3.1.5. The Lactate Pathway

Another proposed pathway that has not yet been constructed in host cells is through a lactate intermediate. However, this pathway was found to be thermodynamically unfavorable [48]. Here, the predicted conversion of lactate leads to 3HP through lactate-CoA, acrylyl-CoA, and 3-hydroxypropionyl-CoA by the action of CoA transferase, lactyl-CoA dehydratase, and 3-hydroxypropionyl-CoA dehydratase, respectively. Finally, conversion of 3-hydroxypropionyl-CoA to 3HP can be performed by one of the following three enzymes: 3-hydroxypropionyl-CoA hydrolase, CoA transferase, or 3-hydroxyisobutyryl-CoA hydrolase.

3.1.6. Use of a Glycerol Intermediate

Another option to convert glucose (or other sugars) to 3HP is by initially producing glycerol from the central metabolism, and finally, converting the glycerol to 3HP with strategies that have been established for glycerol, instead of directly converting the glucose [53].

3.2. Use of E. coli as a Host Microorganism

As discussed previously, *E. coli* is a very commonly used microorganism when it comes to metabolic engineering for the construction of a microbial cell factory, and it has been used extensively for metabolic engineering aimed at 3HP production. On the other hand, the lack of the ability to naturally produce coenzyme B_{12}, which is required for the action of some enzymes involved in the pathways of glycerol conversion to 3HP, is an important drawback. On the other hand, production of 3HP using the malonyl-CoA pathway is coenzyme B_{12}-independent, and presents a promising alternative when using *E. coli* cell factories. As discussed previously, the malonyl-CoA pathway starts with the conversion of acetyl-CoA to malonyl-CoA by the enzyme acetyl-CoA carboxylase, and optimal expression and activity of this enzyme is important. Cheng et al. [54] constructed this route in *E. coli* by expressing the acetyl-CoA carboxylase from *Corynebacterium glutamicum*, and the malonyl-CoA reductase from *Chloroflexus aurantiacus*. To improve the function of the acetyl-CoA carboxylase, the authors added biotin and $NaHCO_3$ (to provide CO_2 for the carboxylation of acetyl-CoA). Under optimal conditions, 1.8 g/L 3HP was produced in flasks with a yield of 0.18 g/g. Scale-up of the process in a fed-batch bioreactor improved the titers obtained to 10.08 g/L (Table 4). It was proposed that low yields of 3HP through the malonyl-CoA pathway can be caused by functional imbalance of the two fragments of the malonyl-CoA reductase (*mcr*), namely the *mcr-c* and *mcr-n* fragments [55]. The *mcr* catalyzes the conversion of malonyl-CoA to 3HP through malonate semialdehyde, with the *mcr-c* fragment catalyzing the first step and the *mcr-n* fragment catalyzing the second. Liu et al. [55] tried to minimize the functional imbalances between the two fragments by employing directed evolution tools, aiming to enhance the action of the *mcr-c* fragment and then fine-tune the levels of expression of the *mcr-n* fragment. The newly engineered strain performed well during fed-batch cultivation on glucose, reaching a 3HP production of to 40.6 g/L, with a conversion yield of 0.19 g/g.

In another study, the β-alanine pathway was constructed in *E. coli* cells [49]. The authors used a previously developed strain, which was capable of producing β-alanine [56], to construct the pathway towards 3HP. Based on this strain, the authors tested several downstream enzymes for their efficiency regarding 3HP production [49]. Among the different combinations, the most promising genes were the *ydfG* from *E. coli* (encoding a malonic semialdehyde reductase) and the *pa0132* from *Pseudomonas aeruginosa* (encoding a β-alanine pyruvate transaminase). Fine-tuning of other genes related to the pathway (such as overexpression of *ppc* gene, encoding phosphoenolpyruvate carboxylase, and replacement of the native promoter with the strong *tac* promoter for the *sdhC* gene, encoding a succinate dehydrogenase) resulted in the production of 31.1 g/L 3HP, with a yield of 0.423 g/g.

Table 4. Results of 3HP production from glucose using *E. coli* as host microorganism.

Genes Transferred	Culture Conditions	Concentration (g/L)	Productivity (g/L·h)	Ref.
acc from *C. glutamicum* and *mcr* from *C. aurantiacus*	Fed-batch bioreactor 5 L, 2.5 L, pH 7, 37 °C, 1 vvm	10.1	0.28	[54]
AccADBC, *mcr* with enhanced activity of the *mcr-c* fragment and tuning of the expression levels of *mcr-n* fragment	Fed-batch bioreactor 5 L, 2 L, pH 7, 37 °C	40.6	0.56	[55]
pa0132 from *P. aeruginosa*, *ydfG* from *E. coli*, upregulation of *sdhC*, overexpression of *ppc*	Fed-batch bioreactor 6.6 L, 2 L, pH 7, 37 °C, 1 vvm, 200–1000 rpm	31.1	0.63	[49]
dhaB and *dhaR* from *Lactobacillus brevis*, *aldhH* from *Pseudomonas aeruginosa*, *gpd1* and *gpp2* from *S. cerevisiae*, deletion of *ptsG*, *glpK* and *yqhD*, overexpression of *xylR* operon	Fed-batch bioreactor 2.5 L with co-conversion of glucose and xylose, 1 L, 37 °C > 25 °C, 1 vvm, 600–1300 rpm	29.7	0.54	[57]

A different approach to conversion of glucose to 3HP would be to take advantage of the naturally produced glycerol in the central metabolism of the host and construct a pathway for the conversion of glycerol to 3HP. In this way, both glucose and xylose (two of the main sugars of lignocellulosic biomass) can be used and channeled to glycerol through glycolysis and the pentose phosphate pathway, respectively. In order to improve the glycerol production yields, Jung et al. [57] constructed an *E. coli* strain by initially expressing the genes encoding glycerol-3-phosphate dehydrogenase (*gpd1*) and glycerol-3-phosphatase (*gpp2*) from *S. cerevisiae*. To achieve co-utilization of xylose and glucose, the *ptsG* gene (encoding the phosphoenolpyruvate:sugar-transferring system) was deleted, and the *xylR* operon was overexpressed. The new strain was capable of simultaneously consuming glucose and xylose, and produce glycerol with a yield of 0.48 g/g. Finally, when the pathway for the conversion of glycerol to 3HP was constructed, 3HP production of 29.7 g/L was accomplished with an overall yield of 0.36 g/g.

3.3. Use of S. cerevisiae as a Host Microorganism

Regarding the use of glucose as carbon source, the yeast *S. cerevisiae* is a very promising candidate, as it is very robust, is easy to handle, and described as GRAS. Another positive characteristic of this yeast is its high tolerance of low pH values, which can result from accumulation of 3HP [58,59]. To permit conversion of glucose to 3HP, the malonyl-CoA pathway was also constructed in *S. cerevisiae* cells [59]. This was done by expressing multiple copies of the *mcr* gene from *C. aurantiacus* and mutated *acc1* genes. Moreover, in an attempt to improve the supply of acetyl-CoA, the native pyruvate decarboxylase *pdc1* gene and the aldehyde dehydrogenase *ald6* gene were overexpressed together with the expression of the acetyl-CoA synthase gene (acs^{L641P}) from *Salmonella enterica*; these modifications increased the 3HP titer by 80%. Finally, the intracellular NADPH, which is required for the action of the *mcr*, was increased by transferring the NADP-dependent glyceraldehyde-3-phosphate dehydrogenase gene (*gapdh*) from *Clostridium acetobutylicum*. The strain constructed produced 9.8 g/L 3HP with a yield of 13% C-mol/C-mol (Table 5).

In another study with *S. cerevisiae* as host microorganism, the β-alanine pathway—which was identified through metabolic modeling to be the most attractive pathway from an economic point of view—was constructed [48]. De novo β-alanine synthesis was performed by the action of a β-alanine-pyruvate aminotransferase (bapat) from *Bacillus cereus* (*yhxA* gene), which is a newly used bapat enzyme. The authors examined the effect on 3HP production of different combinations of the *yhxA* gene (or another bapat gene from *Pseudomonas putida*) with various 3-hydroxypropionate dehydrogenase (hpdh) and 3-hydroxyisobutyrate dehydrogenase (hibadh), the most efficient combination being *yhxA* with the hpdh from *E. coli* (*ydfG* gene). Different aspartate-1-decarboxylases (*panD*) were also tested, the most promising being the one derived from *Tribolium castaneum*. Finally, improvement of the production yields was achieved by improving the supply of L-aspartate by overexpressing a native cytoplasmic aspartate aminotransferase (*aat2*) and two pyruvate carboxylases (*pyc1, pyc2* genes). The strain that was constructed performed well in fed-batch bioreactors, producing 13.7 g/L 3HP with a conversion yield of 0.14 C-mol/C-mol.

Apart from glucose, xylose is also an important sugar that can be derived from lignocellulosic biomass, and it would be important to incorporate it in the 3HP production process. One attempt to convert xylose to 3HP in *S. cerevisiae* was conducted by Kildegaard et al. [60] by incorporating either the malonyl-CoA pathway or the β-alanine pathway (using two approaches, one NADH-dependent route and one NADPH-dependent route) in a xylose-utilizing strain. The three different pathways were tested with the yeast growing on glucose or xylose, and the malonyl-CoA route gave the best results on glucose, and the NADPH-dependent β-alanine pathway resulted in the highest 3HP yields from xylose. The second route was constructed by expressing two pyruvate carboxylases (*pyc1* and *pyc2*), a bapat from *B. cereus*, a *panD* gene from *T. castaneum*, and the NADPH-dependent hpdh from *E. coli* (*ydfG* gene). The strain performed well on xylose, producing 1.8 g/L during flask culture, which increased to 6.1 g/L when the culture was performed in batch bioreactors. Fed-batch cultivation further improved 3HP production to 7.4 g/L, with an overall yield of 29% C-mol/C-mol. Although the concentration was relatively low, the results of this work are very promising, as glucose and xylose are two major sugars in lignocellulosic biomass, and further optimization of the strain would enable culture on lignocellulosic hydrolysates.

Table 5. Results of 3HP production from glucose using *S. cerevisiae* as host microorganism.

Genes Transferred	Culture Conditions	Concentration (g/L)	Productivity (g/L·h)	Ref.
Multiple copies of *mcr* from *C. aurantiacus* and mutant *acc1*, *acs*[L641P] from *S. enterica*, overexpression of *pdc1* and *ald6*, *gapdh* from *C. acetobutylicum*	Fed-batch bioreactor 1 L, 0.5 L, pH 5, 30 °C, 2 vvm, 800 rpm	9.8	0.1	[59]
yhxA from *B. cereus*, *ydfG* from *E. coli*, *panD* from *T. castaneum*, overexpression of *aat2*, *pyc1*, and *pyc2*	Fed-batch bioreactor 1 L, 0.5 L, pH 5, 30 °C, 2 vvm, 800 rpm	13.7	0.17	[48]
pyc1 and *pyc2*, *bapat* from *B. cereus*, *panD* from *T. castaneum*, *ydfG* from *E. coli*	Fed-batch bioreactor 2.7 L on xylose, 1 L, pH 5, 30 °C, 1 vvm, 600–1200 rpm	7.4	0.06	[60]

3.4. Use of Other Host Microorganisms

Apart from work with the two commonly used microorganisms (*E. coli* and *S. cerevisiae*) for the construction of a strain capable of transforming glucose (and in some cases xylose) to 3HP, a few other host microorganisms have also been used. The yeast *Schizosaccharomyces pombe* is one example; this microorganism showed good tolerance of low pH values, which can result the accumulation of 3HP [58]. The malonyl-CoA pathway was constructed in this yeast by overexpression of an endogenous acetyl-CoA carboxylase (*cut6p* gene) and heterologous expression of the *mcr* gene from *C. aurantiacus* in a protease-deficient strain (aimed at protecting the secreted proteins from degradation). Supplementation of the culture broth with acetate was also beneficial for 3HP production; the 3HP concentration reached 3.5 g/L (with a productivity of 0.03 g/L·h) during culture in flasks, which was further increased to 7.6 g/L (with a productivity of 0.25 g/L·h) when high-density culture was used (Table 6).

Table 6. Results of 3HP production from glucose using other host microorganisms.

Genes Transferred	Microorganism	Culture Conditions	Concentration (g/L)	Productivity (g/L·h)	Ref.
mcr from *C. aurantiacus* and overexpression of *cut6p*	*S. pombe*	Flasks 100 mL with high-density cultures, 10 mL, 30 °C, 250 rpm	7.6	0.25	[58]
pdu from *K. pneumoniae*, *gpd1* and *gpp2* from *S. cerevisiae*, *gabD* from *C. necator*, *xylAB* from *E. coli*, *araE* from *C. glutamicum*, deletion of *ldhA*, *pta-ackA*, *poxB*, and *glpK*, and replacement of *ptsH* with *iolT1* and *glk*	*C. glutamicum*	Fed-batch bioreactor 5 L, 2 L, pH 7.2, 30 °C, 1 vvm	62.6	0.87	[53]
pdu from *K. pneumoniae*, *gpd1* and *gpp2* from *S. cerevisiae*, *gabD* from *C. necator*, *xylAB* from *E. coli*, *araE* from *C. glutamicum*, deletion of *ldhA*, *pta-ackA*, *poxB*, and *glpK*, and replacement of *ptsH* with *iolT1* and *glk*	*C. glutamicum*	Fed-batch bioreactor 5 L with glucose and xylose, 2 L, pH 7.2, 30 °C, 1 vvm	54.8	n.a.	[53]

n.a.: not available.

Another host microorganism that was used is the bacterium *Corynebacterium glutamicum*. Chen et al. [53] engineered a strain that could convert glucose and xylose to 3HP through glycerol as an intermediate. The pathway was constructed by using the *pdu* operon from *K. pneumonia*, fusion of the *gpd1* and *gpp2* genes (which are involved in glycerol synthesis), and expression of the *gabD* aldehyde dehydrogenase (which was found to be the most efficient of the aldehyde dehydrogenases tested). Furthermore, genes involved in the formation of by-products were deleted in order to improve the carbon flux towards 3HP. The *ptsH* gene (encoding the phosphoenolpyruvate dependent phosphotransferase system, which is the main route for glucose uptake) was also replaced with the inositol permeases gene (*iolT1*) and glucokinase gene (*glk*); finally, to allow pentose utilization, the pentose transport genes *araE* and the xylose catabolite gene *xylAB* were also expressed. The resulting strain produced 3HP at 37.4 g/L and 35.4 g/L when grown on glucose and xylose, respectively, in batch bioreactors. The amount of 3HP produced for the co-consumption of the two sugars was 36.2 g/L, with a conversion yield of 0.45 g/g. Cultivation in fed-batch bioreactors further improved the 3HP production to 62.6 g/L (0.51 g/g) when growing only on glucose, and to 54.8 g/L (0.49 g/g) when growing on a mixture of glucose and xylose.

4. Production of 3HP from Other Sources

One of the alternative carbon sources that have been used for the production of 3HP is propionic acid. For this reason, the propionyl-CoA pathway (which has already been described) was constructed in *E. coli* by transferring a propionyl-CoA dehydrogenase gene (*pacd*) from *Candida rugora*, a propionate-CoA transferase gene (*pct*) from *Megasphaera elsdenii*, and a 3-hydroxypropionyl-CoA dehydratase gene (*hpcd*) from *C. aurantiacus* [51]. In addition, deletion of the *ygfH* and *prpC* genes led to the production of 3HP at 2.2 g/L (Table 7). The main drawback of using propionate is that it is not a low-cost carbon source, which affects the production cost of 3HP. Moreover, the conversion yield obtained in this work was relatively low (35.4%). On the other hand, as the authors also suggested, more work should be conducted to fully understand and control the pathway, and propionyl-CoA could be derived from glucose through succinyl-CoA, thus avoiding the use of propionic acid.

Recent research has shown that 3HP can also be produced from hydrolysis of the toxic compound 3-hydroxypropionitrile, which is a platform chemical widely used in many organic syntheses for the production of medicines, pesticides, and polymeric compounds [61]. Hydrolysis of 3-hydroxypropionitrile can yield 3HP; however, the chemical hydrolysis is not an environmentally friendly process, as it requires the use of strong basic solutions and high temperatures, and generates hypersaline waste water [62]. Green methods for the hydrolysis of 3-hydroxypropionitrile involve the use of biocatalysts, such as enzymes or whole cells. Zhang et al. [63] tried to isolate microorganisms that had nitrile-hydrolyzing activity from environmental samples. During their work, they identified a yeast strain, *Meyerozyma guilliermondii* CGMCC12935, with nitrile-hydrolyzing activity. The catalytic activity of the resting yeast cells was found to be optimal at 55 °C and at pH 7.5, with the enzyme showing a broad activity against various nitriles (with 3-hydroxypropionitrile being among them). The authors also evaluated the effects of several metal ions and alcohols on the nitrilase activity, and the presence of Ag^+ and Pb^{2+} turned out to be inhibitory. During the bioconversion of 3-hydroxypropionitrile to 3HP from the resting yeast cells, addition of glucose was found to be beneficial for the conversion yield, which reached a production of 19.5 g/L 3HP from 35.5 g/L 3-hydroxypropionitrile. In another study, whole-cell biocatalysis was achieved by using recombinant *E. coli* cells harboring a particular nitrilase gene, which was selected among 15 different nitrilase genes tested [62]. The authors tested both free and immobilized cells, optimizing the incubation parameters for the highest bioconversion yields. The newly developed biocatalyst was very robust and capable of hydrolyzing up to approximately 320 g/L 3-hydroxypropionitrile in 24 h when free cells were used. Immobilization of the cells allowed the use of a concentration of 3-hydroxypropionitrile as high as 497 g/L, which gave complete hydrolysis within 24 h. The immobilized cells showed remarkable stability during the reusability trials, retaining their activity for 30 batches, with a production of 184.7 g/L 3HP during these 30 batches, and a volumetric productivity of 36.9 g/L·h.

Despite the obvious focus on heterotrophic 3HP production from renewable resources, the interest of some research groups lies in studying autotrophic producers, such as photosynthetic cyanobacteria, which have the ability to transform sunlight and CO_2 into a variety of compounds. After modifications by genetic engineering, the recombinant strains are also capable of producing 3HP through CO_2 fixation by the Calvin cycle, and further transformation to phosphoenolpyruvate, which can be converted to 3HP by two newly developed pathways, namely the malonyl-CoA intermediate pathway and the β-alanine intermediate pathway [64,65].

Table 7. Results of 3HP production from carbon sources other than glycerol and glucose (sugars).

Genes Transferred	Microorganism	Culture Conditions	Concentration (g/L)	Productivity (g/L·h)	Ref.
pacd from *C. rugora*, *pct* from *M. elsdenii* and *hpcd* from *C. aurantiacus*, and deletion of *ygfH* and *prpC*	*E. coli*	Flasks 100 mL, 20 mL, pH 7, 30 °C, 200 rpm. Propionic acid as carbon source	2.2	n.a.	[51]
-	*M. guilliermondii*	Flasks, 20 mL, pH 7.2, 30 °C, 120 rpm. Hydrolysis of 3-hydroxypropionitrile with the addition of glucose	19.5	0.20	[63]
Nitrilase from environmental samples	*E. coli*	Batch reaction with immobilized cells with 3-hydroxypropionitrile, 30 °C	184.7	36.9	[62]
mcr from *C. aurantiacus*, overexpression of *pntAB* and *accBCAD-birA*, and deletion of *phaB* and *pta*	*Synechocystis* sp.	Flasks 100 mL, 20 mL, pH 7.5, 30 °C, 150 rpm. From CO_2	0.84	0.006	[64]
mcr from *S. tokodaii* and *msr* from *M. sedula*	*S. elongates*	Flasks, 50 mL, 30 °C. From CO_2	0.67	0.002	[65]
Ppc and *aspC* from *E. coli*, *SkPYD4* from *S. kluyveri*, and *msr* from *M. sedula*	*S. elongates*	Flasks, 50 mL, 30 °C. From CO_2	0.19	0.0005	[65]

n.a.: not available.

The first route involves the use of malonyl-CoA. In this case, phosphoenolpyruvate is transformed through pyruvate to acetyl-CoA, which is converted to malonyl-CoA by the action of acetyl-CoA carboxylase, a biotin-dependent enzyme. Finally, malonyl-CoA is converted to 3HP by malonyl-CoA reductase [64]. This pathway was constructed in the cyanobacterium *Synechocystis* sp. by transferring the *mcr* gene from *C. aurantiacus* [64]. To improve the supply of NADPH and malonyl-CoA, the genes for NAD(P)-transhydrogenase (*pntAB*) and acetyl-CoA carboxylase and biotinilase (*accBCAD-birA*) were overexpressed. Finally, to improve the carbon flux towards 3HP, the genes involved in the formation of PHA (*phaB*) and acetate (*pta*) were deleted. The newly constructed strain was capable of producing approximately 0.84 g/L 3HP directly from CO_2. The malonyl-CoA route was also used by Lan et al. [65] for the construction of a recombinant *Synechococcus elongatus* strain. The authors expressed the *mcr* gene from *Sulfolobus tokodaii* and the malonate semialdehyde reductase gene (*msr*) from *Metallosphaera sedula*, which resulted in the production of 0.67 g/L of 3HP.

The second approach is to use the β-alanine intermediate pathway, where phosphoenolpyruvate is converted to oxaloacetate by the action of phosphoenolpyruvate carboxylase. Oxaloacetate is transformed by the action of aspartate transaminase to aspartate, which is then converted to β-alanine by the action of aspartate carboxylase. Finally, β-alanine is converted to malonate semialdehyde by the action of β-alanine aminotransferase, and finally, to 3HP by the action of malonate semialdehyde reductase [65]. Lan et al. [65] constructed this pathway in *S. elongates* cells by expressing the phosphoenolpyruvate carboxylase gene (*ppc*) and the aspartate aminotransferase gene (*aspC*) from *E. coli*, together with the β-alanine aminotransferase gene (*SkPYD4*) from *Saccharomyces kluyveri* and the *msr* gene from *M. sedula*. Production of 3HP from CO_2 was approximately 0.19 g/L, which was lower than the results obtained when 3HP was produced through the malonyl-CoA intermediate pathway. Although the results obtained from the direct conversion of CO_2 to 3HP are still low, direct conversion of CO_2 to high-value chemicals is a very promising process, and during the next year, more research work is expected to be devoted to this, with the aim of improving the conversion yields.

5. Conclusions

Biotechnological production of 3HP from renewable resources, the treatment of toxic compounds, and CO_2 have been attracting considerable attention during the decades. Different routes and techniques are being established, with the goal of a sustainable fossil-free production of 3HP. During the last four years, significant progress has been made in improving the production yields of 3HP from glycerol and glucose, which has resulted in improvement of the titers achieved, and also, the conversion yields. New approaches have also been proposed, such as systems for the hydrolysis of 3-hydroxypropionitrile, resulting in high concentrations of 3HP with concomitant treatment of a toxic compound. The very good results that have been obtained through all the research work conducted are an important step towards the commercial biotechnological production of 3HP.

Acknowledgments: We thank Bio4Energy, a strategic research environment appointed by the Swedish government, for supporting this work.

Author Contributions: Leonidas Matsakas, Ulrika Rova, and Paul Christakopoulos conceived the article and planned the structure of the manuscript. Leonidas Matsakas and Kateřina Hrůzová searched and analyzed the literature and wrote the paper. All the authors have read, commented on, and approved the paper.

Conflicts of Interest: The authors declare that they have no conflicts of interest.

References

1. Werpy, T.; Petersen, G.; Aden, A.; Bozell, J.; Holladay, J.; White, J.; Manheim, A.; Eliot, D.; Lasure, L.; Jones, S. *Top Value Added Chemicals from Biomass. Volume 1-Results of Screening for Potential Candidates from Sugars and Synthesis Gas*; Department of Energy: Washington, DC, USA, 2004.

2. Rathnasingh, C.; Raj, S.M.; Jo, J.E.; Park, S. Development and evaluation of efficient recombinant *Escherichia coli* strains for the production of 3-hydroxypropionic acid from glycerol. *Biotechnol. Bioeng.* **2009**, *104*, 729–739. [CrossRef] [PubMed]

3. Matsakas, L.; Topakas, E.; Christakopoulos, P. New trends in microbial production of 3-hydroxypropionic acid. *Curr. Biochem. Eng.* **2015**, *1*, 141–154. [CrossRef]

4. Jung, W.S.; Kang, J.H.; Chu, H.S.; Choi, I.S.; Cho, K.M. Elevated production of 3-hydroxypropionic acid by metabolic engineering of the glycerol metabolism in *Escherichia coli*. *Metab. Eng.* **2014**, *23*, 116–122. [CrossRef] [PubMed]

5. Matsakas, L.; Giannakou, M.; Vörös, D. Effect of synthetic and natural media on lipid production from *Fusarium oxysporum*. *Electron. J. Biotechnol.* **2017**, *30*, 95–102. [CrossRef]

6. Yang, F.; Hanna, M.A.; Sun, R. Value-added uses for crude glycerol—A byproduct of biodiesel production. *Biotechnol. Biofuels* **2012**, *5*, 13. [CrossRef] [PubMed]

7. Okoye, P.U.; Hameed, B.H. Review on recent progress in catalytic carboxylation and acetylation of glycerol as a byproduct of biodiesel production. *Renew. Sustain. Energy Rev.* **2016**, *53*, 558–574. [CrossRef]

8. He, Q.; McNutt, J.; Yang, J. Utilization of the residual glycerol from biodiesel production for renewable energy generation. *Renew. Sustain. Energy Rev.* **2017**, *71*, 63–76. [CrossRef]

9. Rodrigues, A.; Bordado, J.C.; Dos Santos, R.G. Upgrading the glycerol from biodiesel production as a source of energy carriers and chemicals—A technological review for three chemical pathways. *Energies* **2017**, *10*, 1817. [CrossRef]

10. Garlapati, V.K.; Shankar, U.; Budhiraja, A. Bioconversion technologies of crude glycerol to value added industrial products. *Biotechnol. Rep.* **2016**, *9*, 9–14. [CrossRef] [PubMed]

11. Ashok, S.; Raj, S.M.; Rathnasingh, C.; Park, S. Development of recombinant *Klebsiella pneumoniae* Δ*dhaT* strain for the co-production of 3-hydroxypropionic acid and 1,3-propanediol from glycerol. *Appl. Microbiol. Biotechnol.* **2011**, *90*, 1253–1265. [CrossRef] [PubMed]

12. Ko, Y.; Seol, E.; Sundara Sekar, B.; Kwon, S.; Lee, J.; Park, S. Metabolic engineering of *Klebsiella pneumoniae* J2B for co-production of 3-hydroxypropionic acid and 1,3-propanediol from glycerol: Reduction of acetate and other by-products. *Bioresour. Technol.* **2017**, *244*, 1096–1103. [CrossRef] [PubMed]

13. Chen, X.; Xiu, Z.; Wang, J.; Zhang, D.; Xu, P. Stoichiometric analysis and experimental investigation of glycerol bioconversion to 1,3-propanediol by *Klebsiella pneumoniae* under microaerobic conditions. *Enzyme Microb. Technol.* **2003**, *33*, 386–394. [CrossRef]

14. Zaushitsyna, O.; Dishisha, T.; Hatti-Kaul, R.; Mattiasson, B. Crosslinked, cryostructured *Lactobacillus reuteri* monoliths for production of 3-hydroxypropionaldehyde, 3-hydroxypropionic acid and 1,3-propanediol from glycerol. *J. Biotechnol.* **2017**, *241*, 22–32. [CrossRef] [PubMed]

15. Ko, Y.; Ashok, S.; Ainala, S.K.; Sankaranarayanan, M.; Chun, A.Y.; Jung, G.Y.; Park, S. Coenzyme B_{12} can be produced by engineered *Escherichia coli* under both anaerobic and aerobic conditions. *Biotechnol. J.* **2014**, *9*, 1526–1535. [CrossRef] [PubMed]

16. Lawrence, J.G.; Roth, J.R. Evolution of coenzyme B_{12} synthesis among enteric bacteria: Evidence for loss and reacquisition of a multigene complex. *Genetics* **1996**, *142*, 11–24. [PubMed]

17. Toraya, T. Radical catalysis in coenzyme B_{12}-dependent isomerization (eliminating) reactions. *Chem. Rev.* **2003**, *103*, 2095–2127. [CrossRef] [PubMed]

18. Daniel, R.; Bobik, T.A.; Gottschalk, G. Biochemistry of coenzyme B12-dependent glycerol and diol dehydratases and organization of the encoding genes. *FEMS Microbiol. Rev.* **1998**, *22*, 553–566. [CrossRef] [PubMed]

19. Zhang, G.L.; Xu, X.L.; Li, C.; Ma, B. Bin Cloning, expression and reactivating characterization of glycerol dehydratase reactivation factor from *Klebsiella pneumoniae* XJPD-Li. *World J. Microbiol. Biotechnol.* **2009**, *25*, 1947–1953. [CrossRef]

20. Ashok, S.; Sankaranarayanan, M.; Ko, Y.; Jae, K.E.; Ainala, S.K.; Kumar, V.; Park, S. Production of 3-hydroxypropionic acid from glycerol by recombinant *Klebsiella pneumoniae* $\Delta dhaT\Delta yqhD$ which can produce vitamin B_{12} naturally. *Biotechnol. Bioeng.* **2013**, *110*, 511–524. [CrossRef] [PubMed]

21. Luo, L.H.; Seo, J.W.; Baek, J.O.; Oh, B.R.; Heo, S.Y.; Hong, W.K.; Kim, D.H.; Kim, C.H. Identification and characterization of the propanediol utilization protein PduP of *Lactobacillus reuteri* for 3-hydroxypropionic acid production from glycerol. *Appl. Microbiol. Biotechnol.* **2011**, *89*, 697–703. [CrossRef] [PubMed]

22. Bobik, T.A.; Havemann, G.D.; Busch, R.J.; Williams, D.S.; Aldrich, H.C. The propanediol utilization (*pdu*) operon of *Salmonella enterica* serovar Typhimurium LT2 includes genes necessary for formation of polyhedral organelles involved in coenzyme B_{12}-dependent 1,2-propanediol degradation. *J. Bacteriol.* **1999**, *181*, 5967–5975. [PubMed]

23. Dishisha, T.; Pereyra, L.P.; Pyo, S.H.; Britton, R.A.; Hatti-Kaul, R. Flux analysis of the *Lactobacillus reuteri* propanediol-utilization pathway for production of 3-hydroxypropionaldehyde, 3-hydroxypropionic acid and 1,3-propanediol from glycerol. *Microb. Cell Fact.* **2014**, *13*. [CrossRef] [PubMed]

24. Johnsont, E.A.; Lin, E.C.C. *Klebsiella pneumoniae* 1,3-Propanediol:NAD^{+} Oxidoreductase. *J. Bacteriol.* **1987**, *169*, 2050–2054. [CrossRef]

25. Wang, K.; Tian, P. Engineering plasmid-free *Klebsiella pneumoniae* for production of 3-hydroxypropionic acid. *Curr. Microbiol.* **2017**, *74*, 55–58. [CrossRef] [PubMed]

26. Podschun, R.; Ullmann, U. *Klebsiella* spp. as nosocomial pathogens: Epidemiology, taxonomy, typing methods, and pathogenicity factors. *Clin. Microbiol. Rev.* **1998**, *11*, 589–603. [PubMed]

27. Su, M.; Li, Y.; Ge, X.; Tian, P. 3-Hydroxypropionaldehyde-specific aldehyde dehydrogenase from *Bacillus subtilis* catalyzes 3-hydroxypropionic acid production in *Klebsiella pneumoniae*. *Biotechnol. Lett.* **2015**, *37*, 717–724. [CrossRef] [PubMed]

28. Huang, Y.; Li, Z.; Shimizu, K.; Ye, Q. Co-production of 3-hydroxypropionic acid and 1,3-propanediol by *Klebseilla pneumoniae* expressing *aldH* under microaerobic conditions. *Bioresour. Technol.* **2013**, *128*, 505–512. [CrossRef] [PubMed]

29. Li, Y.; Wang, X.; Ge, X.; Tian, P. High production of 3-hydroxypropionic acid in *Klebsiella pneumoniae* by systematic optimization of glycerol metabolism. *Sci. Rep.* **2016**, *6*, 26932. [CrossRef] [PubMed]

30. Zhao, L.; Lin, J.; Wang, H.; Xie, J.; Wei, D. Development of a two-step process for production of 3-hydroxypropionic acid from glycerol using *Klebsiella pneumoniae* and *Gluconobacter oxydans*. *Bioprocess Biosyst. Eng.* **2015**, *38*, 2487–2495. [CrossRef] [PubMed]

31. Niu, K.; Cheng, X.L.; Qin, H.B.; Liu, J.S.; Zheng, Y.G. Investigation of the key factors on 3-hydroxypropionic acid production with different recombinant strains. *3 Biotech* **2017**, *7*, 314. [CrossRef] [PubMed]

32. Kumar, V.; Ashok, S.; Park, S. Recent advances in biological production of 3-hydroxypropionic acid. *Biotechnol. Adv.* **2013**, *31*, 945–961. [CrossRef] [PubMed]

33. Lim, H.G.; Noh, M.H.; Jeong, J.H.; Park, S.; Jung, G.Y. Optimum rebalancing of the 3-hydroxypropionic acid production pathway from glycerol in *Escherichia coli*. *ACS Synth. Biol.* **2016**, *5*, 1247–1255. [CrossRef] [PubMed]

34. Chu, H.S.; Kim, Y.S.; Lee, C.M.; Lee, J.H.; Jung, W.S.; Ahn, J.H.; Song, S.H.; Choi, I.S.; Cho, K.M. Metabolic engineering of 3-hydroxypropionic acid biosynthesis in *Escherichia coli*. *Biotechnol. Bioeng.* **2015**, *112*, 356–364. [CrossRef] [PubMed]

35. Sankaranarayanan, M.; Somasundar, A.; Seol, E.; Chauhan, A.S.; Kwon, S.; Jung, G.Y.; Park, S. Production of 3-hydroxypropionic acid by balancing the pathway enzymes using synthetic cassette architecture. *J. Biotechnol.* **2017**, *259*, 140–147. [CrossRef] [PubMed]

36. Niu, K.; Xiong, T.; Qin, H.B.; Wu, H.; Liu, Z.Q.; Zheng, Y.G. 3-Hydroxypropionic acid production by recombinant *Escherichia coli* ZJU-3HP01 using glycerol–glucose dual-substrate fermentative strategy. *Biotechnol. Appl. Biochem.* **2017**, *64*, 572–578. [CrossRef] [PubMed]

37. Sankaranarayanan, M.; Ashok, S.; Park, S. Production of 3-hydroxypropionic acid from glycerol by acid tolerant *Escherichia coli. J. Ind. Microbiol. Biotechnol.* **2014**, *41*, 1039–1050. [CrossRef] [PubMed]

38. Tsuruno, K.; Honjo, H.; Hanai, T. Enhancement of 3-hydroxypropionic acid production from glycerol by using a metabolic toggle switch. *Microb. Cell Fact.* **2015**, *14*, 155. [CrossRef] [PubMed]

39. Honjo, H.; Tsuruno, K.; Tatsuke, T.; Sato, M.; Hanai, T. Dual synthetic pathway for 3-hydroxypropionic acid production in engineered *Escherichia coli. J. Biosci. Bioeng.* **2015**, *120*, 199–204. [CrossRef] [PubMed]

40. Amin, H.M.; Hashem, A.M.; Ashour, M.S.; Hatti-Kaul, R. 1,2 Propanediol utilization by *Lactobacillus reuteri* DSM 20016, role in bioconversion of glycerol to 1,3 propanediol, 3-hydroxypropionaldehyde and 3-hydroxypropionic acid. *J. Genet. Eng. Biotechnol.* **2013**, *11*, 53–59. [CrossRef]

41. Dishisha, T.; Pyo, S.H.; Hatti-Kaul, R. Bio-based 3-hydroxypropionic- and acrylic acid production from biodiesel glycerol via integrated microbial and chemical catalysis. *Microb. Cell Fact.* **2015**, *14*, 200. [CrossRef] [PubMed]

42. Kalantari, A.; Chen, T.; Ji, B.; Stancik, I.A.; Ravikumar, V.; Franjevic, D.; Saulou-Bérion, C.; Goelzer, A.; Mijakovic, I. Conversion of glycerol to 3-hydroxypropanoic acid by genetically engineered *Bacillus subtilis. Front. Microbiol.* **2017**, *8*, 638. [CrossRef] [PubMed]

43. Sjöblom, M.; Matsakas, L.; Krige, A.; Rova, U.; Christakopoulos, P. Direct electricity generation from sweet sorghum stalks and anaerobic sludge. *Ind. Crops Prod.* **2017**, *108*, 505–511. [CrossRef]

44. Matsakas, L.; Nitsos, C.; Vörös, D.; Rova, U.; Christakopoulos, P. High-titer methane from organosolv-pretreated spruce and birch. *Energies* **2017**, *10*, 263. [CrossRef]

45. Nitsos, C.; Matsakas, L.; Triantafyllidis, K.; Rova, U.; Christakopoulos, P. Investigation of different pretreatment methods of Mediterranean-type ecosystem agricultural residues: Characterisation of pretreatment products, high-solids enzymatic hydrolysis and bioethanol production. *Biofuels* **2017**, 1–14. [CrossRef]

46. Matsakas, L.; Rova, U.; Christakopoulos, P. Strategies for enhanced biogas generation through anaerobic digestion of forest material—An overview. *BioResources* **2016**, *11*, 5482–5499. [CrossRef]

47. Choi, S.; Song, C.W.; Shin, J.H.; Lee, S.Y. Biorefineries for the production of top building block chemicals and their derivatives. *Metab. Eng.* **2015**, *28*, 223–239. [CrossRef] [PubMed]

48. Borodina, I.; Kildegaard, K.R.; Jensen, N.B.; Blicher, T.H.; Maury, J.; Sherstyk, S.; Schneider, K.; Lamosa, P.; Herrgård, M.J.; Rosenstand, I.; et al. Establishing a synthetic pathway for high-level production of 3-hydroxypropionic acid in *Saccharomyces cerevisiae* via β-alanine. *Metab. Eng.* **2015**, *27*, 57–64. [CrossRef] [PubMed]

49. Song, C.W.; Kim, J.W.; Cho, I.J.; Lee, S.Y. Metabolic engineering of *Escherichia coli* for the production of 3-hydroxypropionic acid and malonic acid through β-alanine route. *ACS Synth. Biol.* **2016**, *5*, 1256–1263. [CrossRef] [PubMed]

50. Hasegawa, J.; Ogura, M.; Kanema, H.; Kawaharada, H.; Watanabe, K. Production of β-hydroxypropionic acid from propionic acid by a *Candida rugosa* mutant unable to assimilate propionic acid: Studies on β-hydroxycarboxylic acids (IV). *J. Ferment. Technol.* **1982**, *60*, 591–594.

51. Luo, H.; Zhou, D.; Liu, X.; Nie, Z.; Quiroga-Sánchez, D.L.; Chang, Y. Production of 3-hydroxypropionic acid via the propionyl-CoA pathway using recombinant *Escherichia coli* strains. *PLoS ONE* **2016**, *11*, e0156286. [CrossRef] [PubMed]

52. Burgard, A.P.; Dien van, S.J. Methods and Organisms for Growth-Coupled Production of 3-Hydroxypropionic Acid. U.S. Patent 2008/0199926 A1, 22 January 2008.

53. Chen, Z.; Huang, J.; Wu, Y.; Wu, W.; Zhang, Y.; Liu, D. Metabolic engineering of *Corynebacterium glutamicum* for the production of 3-hydroxypropionic acid from glucose and xylose. *Metab. Eng.* **2017**, *39*, 151–158. [CrossRef] [PubMed]

54. Cheng, Z.; Jiang, J.; Wu, H.; Li, Z.; Ye, Q. Enhanced production of 3-hydroxypropionic acid from glucose via malonyl-CoA pathway by engineered *Escherichia coli*. *Bioresour. Technol.* **2016**, *200*, 897–904. [CrossRef] [PubMed]

55. Liu, C.; Ding, Y.; Zhang, R.; Liu, H.; Xian, M.; Zhao, G. Functional balance between enzymes in malonyl-CoA pathway for 3-hydroxypropionate biosynthesis. *Metab. Eng.* **2016**, *34*, 104–111. [CrossRef] [PubMed]

56. Song, C.W.; Lee, J.; Ko, Y.S.; Lee, S.Y. Metabolic engineering of *Escherichia coli* for the production of 3-aminopropionic acid. *Metab. Eng.* **2015**, *30*, 121–129. [CrossRef] [PubMed]

57. Jung, I.Y.; Lee, J.W.; Min, W.K.; Park, Y.C.; Seo, J.H. Simultaneous conversion of glucose and xylose to 3-hydroxypropionic acid in engineered *Escherichia coli* by modulation of sugar transport and glycerol synthesis. *Bioresour. Technol.* **2015**, *198*, 709–716. [CrossRef] [PubMed]

58. Suyama, A.; Higuchi, Y.; Urushihara, M.; Maeda, Y.; Takegawa, K. Production of 3-hydroxypropionic acid via the malonyl-CoA pathway using recombinant fission yeast strains. *J. Biosci. Bioeng.* **2017**, *124*, 392–399. [CrossRef] [PubMed]

59. Kildegaard, K.R.; Jensen, N.B.; Schneider, K.; Czarnotta, E.; Özdemir, E.; Klein, T.; Maury, J.; Ebert, B.E.; Christensen, H.B.; Chen, Y.; et al. Engineering and systems-level analysis of *Saccharomyces cerevisiae* for production of 3-hydroxypropionic acid via malonyl-CoA reductase-dependent pathway. *Microb. Cell Fact.* **2016**, *15*, 53. [CrossRef] [PubMed]

60. Kildegaard, K.R.; Wang, Z.; Chen, Y.; Nielsen, J.; Borodina, I. Production of 3-hydroxypropionic acid from glucose and xylose by metabolically engineered *Saccharomyces cerevisiae*. *Metab. Eng. Commun.* **2015**, *2*, 132–136. [CrossRef]

61. Xu, S.; Xiang, A.; Ying, A. Purification of 3-hydroxypropionitrile by wiped molecular distillation. *Sci. China Ser. B* **2004**, *47*, 521. [CrossRef]

62. Yu, S.; Yao, P.; Li, J.; Ren, J.; Yuan, J.; Feng, J.; Wang, M.; Wu, Q.; Zhu, D. Enzymatic synthesis of 3-hydroxypropionic acid at high productivity by using free or immobilized cells of recombinant *Escherichia coli*. *J. Mol. Catal. B Enzym.* **2016**, *129*, 37–42. [CrossRef]

63. Zhang, Q.; Gong, J.S.; Dong, T.T.; Liu, T.T.; Li, H.; Dou, W.F.; Lu, Z.M.; Shi, J.S.; Xu, Z.H. Nitrile-hydrolyzing enzyme from *Meyerozyma guilliermondii* and its potential in biosynthesis of 3-hydroxypropionic acid. *Bioprocess Biosyst. Eng.* **2017**, *40*, 901–910. [CrossRef] [PubMed]

64. Wang, Y.; Sun, T.; Gao, X.; Shi, M.; Wu, L.; Chen, L.; Zhang, W. Biosynthesis of platform chemical 3-hydroxypropionic acid (3-HP) directly from CO_2 in cyanobacterium *Synechocystis* sp. PCC 6803. *Metab. Eng.* **2016**, *34*, 60–70. [CrossRef] [PubMed]

65. Lan, E.I.; Chuang, D.S.; Shen, C.R.; Lee, A.M.; Ro, S.Y.; Liao, J.C. Metabolic engineering of cyanobacteria for photosynthetic 3-hydroxypropionic acid production from CO_2 using *Synechococcus elongatus* PCC 7942. *Metab. Eng.* **2015**, *31*, 163–170. [CrossRef] [PubMed]

MDPI

St. Alban-Anlage 66

4052 Basel, Switzerland

Tel. +41 61 683 77 34

Fax +41 61 302 89 18

http://www.mdpi.com

Fermentation Editorial Office

E-mail: fermentation@mdpi.com

http://www.mdpi.com/journal/fermentation

www.ingramcontent.com/pod-product-compliance
Lightning Source LLC
Chambersburg PA
CBHW041219220326
41597CB00033BA/6035